Jeremy Rifkin grew up in Chicago [...] sity of Pennsylvania as an und[...] master's degree in International [...] School of Law and Diplomacy at Tufts University. He has been an activist since 1967, when he was a sponsor of the first national rally in the USA against the Vietnam war. He has appeared on television and lectures regularly to university and community audiences throughout the United States. He has served as an adviser to US Congressional Committees and to labour and management on economic and social issues. He was selected by a Presidential Commission as one of twelve economists to provide expert testimony on future options for the US economy; and by the Congressional Clearinghouse on the Future to provide the first in-depth briefing on genetic engineering for the leadership of the House of Representatives. He is the author of *Common Sense II*, *Own Your Own Job*, *The North Will Rise Again* (with Randy Barber) and has written two other books with Ted Howard: *Who Should Play God?* and *The Emerging Order*. He is also author (with Nicanor Perlas) of *Algeny*. Jeremy Rifkin lives in Washington, DC.

Ted Howard is the author of *Voices of the American Revolution*.

JEREMY RIFKIN
with Ted Howard

Entropy

A New World View

Afterword by Dr Nicholas Georgescu-Roegen

PALADIN
Granada Publishing

Paladin Books
Granada Publishing Ltd
8 Grafton Street, London W1X 3LA

Published by Paladin Books 1985

Grateful acknowledgement is made to Harvard University
Press for permission to reprint an excerpt from *The
Entropy Law and the Economic Process* by Nicholas
Georgescu-Roegen.

ISBN 0-586-08508-4

Reproduced, printed and bound in Great Britain by
Hazell Watson & Viney Limited,
Aylesbury, Bucks

Set in Ehrhardt

To Dr Nicholas Georgescu-Roegen
who laid the groundwork by persevering
as prophet and teacher.

Acknowledgements

Special Editorial Contributions:
Noreen Banks
Dan Smith

Our special thanks to Marylin McDonald and Jeffrey Apter for their help in the copy-editing and preparation of the manuscript.

We would also like to acknowledge the contributions of Jenny Speicher and Randy Barber of the Peoples Business Commission.

We would especially like to thank Alan D. Williams, our editor at The Viking Press; his belief in the importance of this book has been a constant source of encouragement.

Contents

Author's Note

Hope is the feeling that what is desired is possible of attainment. This book is about hope: the hope that comes from shattering false illusions and replacing them with new truths.

To a civilization nurtured on the modernist notion of a future without physical constraints and a world without material boundaries, the truths of the Entropy Law will at first appear sobering, even sombre. That is because this law defines the ultimate physical boundaries within which we are constrained to act.

If we continue to ignore the truth of the Entropy Law and its role in defining the broad context in which our physical world unfolds, then we shall do so at the risk of our own extinction.

After finishing this book some will remain unconvinced that there are physical limits that place restraints on human action in the world. Others will be convinced but will conclude with despair that the Entropy Law is a giant cosmic prison from which there is no escape. Finally, there will be those who will see the Entropy Law as the truth that can set us free. The first group will continue to uphold the existing world paradigm. The second group will be without a world view. The third group will be the harbingers of the new age.

PART ONE
World Views

Introduction

Each day we awake to a world that appears more confused and disordered than the one we left the night before. Nothing seems to work anymore. Our lives are bound up in constant repair. We are forever mending and patching. Our leaders are forever lamenting and apologizing. Every time we think we've found a way out of a crisis, something backfires. The powers that be continue to address the problems at hand with solutions that create even greater problems than the ones they were meant to solve.

There are accidents at nuclear power plants, shootouts in gas station lines over fuel allocations, the doubling and tripling of inflation figures, the steady loss of productivity and jobs, the increased danger of thermonuclear war, and finally we want to roll down the window and scream out in desperation, 'Why isn't something being done about all this!' We blame the oil companies, the government economists, the unions, the intellectuals, and anyone else we can target who's at all involved, and still things get worse.

We look around us only to find that the garbage and pollution are piling up in every quarter, oozing out of the ground, seeping into our rivers, and lingering in our air. Our eyes burn, our skin discolours, our lungs collapse, and all we can think of is retreating indoors and closing the shutters.

Everywhere we go we find ourselves waiting in lines or pushed into corners. Things about us continue to accelerate, yet nothing seems to be getting anywhere. We are bogged down, the society is bogged down; and all of a sudden we get this urge to trample over everything in our path, leaving the world behind us in disarray.

They tell us it's no better anywhere else, and for once they're right. We look at the other industrial societies, and while some appear worse off, and others slightly better off, all of them, socialist and capitalist alike, seem to be gripped by a common malaise. The same inexorable force of disintegration is eating away at us all.

When the whole world begins to break down and fall apart, then we must look to the way the whole world has been organized, because that is where the problem lies. For it is foolish to continue to blame individual leaders and individual ideologies for a universal problem. Of course some leaders and some ideologies are better than others. Still, at the present time no single leader or ideology on this planet can effectively address the universal crisis at hand, because all are committed to the existing world view, one that is diseased and dying and is contaminating everything it gave birth to.

World Views

Throughout history, human beings have felt the need to construct a frame of reference for organizing life's activities. The need to establish an order to explain the hows and whys of daily existence has been the essential cultural ingredient of every society. The most interesting aspect of a society's world view is that its individual adherents are, for the most part, unconscious of how it affects the way they do things and how they perceive the reality around them. A world view is successful to the extent that it is so internalized, from childhood on, that it goes unquestioned.

Most Americans believe that the world is progressing toward a more valuable state as a result of the steady accumulation of human knowledge and techniques. We also believe that the individual exists as an autonomous entity, that nature has an order to it, that scientific observation is objective, that people have always desired private property, that competition between individuals has always occurred, and so on. In fact, all these beliefs are considered to be part of 'human nature' and therefore immutable. Of course, they are not, and other societies and civilizations at other periods in history would simply be unable to comprehend some of the notions we ascribe to human nature. That is the power of a world view. Its hold over our perception of reality is so overwhelming that we can't possibly imagine any other way of looking at the world.

Our modern world view took shape some 400 years ago, and while it has been greatly refined and modified in the years since, it has retained much of its early vision. We live under the influence of the seventeenth-century Newtonian world machine paradigm. In the next chapter we will explore this paradigm in detail. There probably isn't one person in a hundred who could explain the intricate features of Newtonian mechanics; nonetheless, its shadow is always with us, influencing our every move.

Now, however, a new world view is about to emerge, one that will eventually replace the Newtonian world machine as the organizing frame of history: the Entropy Law will preside as the ruling paradigm over the next period of history. Albert Einstein said that it is the premier law of all of science; Sir Arthur Eddington referred to it as the supreme metaphysical law of the entire universe. The Entropy Law is the second law of thermodynamics. The first law states that all matter and energy in the universe is constant, that it cannot be created or destroyed. Only its form can change but never its essence. The second law, the Entropy Law, states that matter and energy can only be changed in one direction, that is, from usable to unusable, or from available to unavailable, or from ordered to disordered. In essence, the second law says that everything in the entire universe began with structure and value and is irrevocably moving in the direction of random chaos and waste. Entropy is a measure of the extent to which available energy in any subsystem of the universe is transformed into an unavailable form. According to the Entropy Law, whenever a semblance of order is created anywhere on earth or in the universe, it is done at the expense of causing an even greater disorder in the surrounding environment. The Entropy Law will be explained in detail in Part 2.

For now a few simple observations are in order, observations that the reader will have to accept on faith, at least until we perform a thorough autopsy on the prevailing world view and explore the hidden dimensions of the new entropy paradigm. The Entropy Law destroys the notion of history as progress. The Entropy Law destroys the notion that science and technology create a more ordered world. In fact, the Entropy Law transcends the modern world view with a force of conviction that is every bit as convincing as was the Newtonian world machine when it replaced the medieval Christian world view of the Roman Church.

Step by step the Entropy Law will provide us with an understanding of exactly why the existing paradigm has broken down. Our generation, caught between the old paradigm that we were nurtured on and the new entropy paradigm just emerging, will begin to marvel at how we could have believed in principles and axioms so obviously false. We will stumble into the new paradigm, ill at ease and groping like a visitor to a foreign land. Unable to completely shed our native world view, we shall take on the new entropy paradigm as a second language, never completely comfortable with it and never able to fully articulate it into our daily routines. For our grandchildren's generation the entropic world view will be like second nature: they will not think about it, they will merely live by it, unconscious of its hold over them, as we have for so long been unconscious of the hold Newtonian mechanics has had over us.

Already the outline of the new entropy paradigm is being filled in by scholars around the world. Within a few years every academic discipline will be turned inside out by the new entropy conception. There will be attempts to graft the Entropy Law onto the existing world view, a task that will ultimately fail. Politicians will proclaim its importance in addressing issues ranging from energy to disarmament. Theologians will construct new interpretations of Biblical authority based on it. Technicians will develop new approaches to problem solving in the misguided belief that it can be quantified and reduced to precise measurement. Economists will scramble to redesign classical economic theory to conform with its central truths. Psychologists and sociologists will reexamine human nature with entropy as a backdrop. All this will happen in the next few years. All this and more. So much more that none of us can even make out more than the shadows and echoes of the world that is being born within us today.

There will also be those who will stubbornly refuse to accept the fact that the Entropy Law reigns supreme over all physical reality in the world. They will insist that the entropy process only applies in selective instances and that any attempt to apply it more broadly to society is to engage in the use of metaphor. Quite simply, they are wrong. The laws of thermodynamics provide the overarching scientific frame for the unfolding of *all physical activity* in this world. In the words of the Nobel Prize-winning chemist Frederick Soddy, the laws of thermodynamics 'control, in the last resort, the rise

17

and fall of political systems, the freedom or bondage of Nations, the movements of commerce and industry, the origins of wealth and poverty, and the general physical welfare of the race.' Every single physical activity that humankind engages in is totally subject to the iron-clad imperative expressed in the first and second laws of thermodynamics.

It should be emphasized that the Entropy Law deals only with the physical world where everything is finite and where all living things must run their course and eventually cease to be. It is a law governing the horizontal world of time and space. It is mute, however, when it comes to the vertical world of spiritual transcendence. The spiritual plane is not governed by the ironclad dictates of the Entropy Law. The spirit is a nonmaterial dimension where there are no boundaries and no fixed limits to attend to. The relationship of the physical to the spiritual world is the relationship of a small part to the larger unbound whole within which it unfolds. While the Entropy Law governs the world of time, space, and matter, it is, in turn, governed by the primordial spiritual force that conceived it.

The way a civilization organizes its physical reality and the importance it attaches to the material plane of existence determine how favourable the conditions are for seeking spiritual enlightenment. The more steeped a world view is in the material side of life, the less conducive it is to the human quest for spiritual transcendence. The less attached a civilization is to the physical world, the freer the human collectivity is to transcend the confines of the material plane and become one with the profound spiritual essence that encompasses all.

The laws of thermodynamics, then, govern the physical world. The way humanity decides to interact with those laws in establishing a framework for physical existence is of crucial importance in whether humankind's spiritual journey is allowed to flourish or languish. A thorough comprehension of the Entropy Law is crucial for understanding the physical context from which all spiritual sojourns must start.

Historians and anthropologists have long speculated over why a particular world view emerges at a particular time and place in history. This essay will suggest an answer to that question: that the energy condition of the environment sets the broad frame for the world view that emerges. But before attempting to demonstrate that claim, it's important that we remove ourselves from our own world view just long enough to take a hard look at how our own perception of reality has been shaped over the centuries.

The Greeks and the Five Ages of History:
Cycles and Decay

Certainly, Plato, Aristotle, and the other Greek philosophers were no fools. How then do we account for the fact that they saw history in a way that's exactly opposite to how we perceive it? For the Greeks, history was a process of continual degradation. Horace, a Roman, mused that 'time depreciates the value of the world.'[1] Horace didn't know about the second law of thermodynamics, but in this verse he summed up the very essence of the Entropy Law (as we will see in Part 2). In Greek mythology, history is represented by a series of five stages, each more degraded and more harsh than the one preceding it. The Greek historian Hesiod describes these ages as the Golden, Silver, Brass, Heroic, and Iron. The Golden Age was the apex: a period of abundance and fulfilment.

In the beginning, a golden race of mortal men was made by the immortal dwellers on Olympus . . . They lived like Gods with hearts free from care, without part or lot in labour and sorrow. Pitiful old age did not await them, but ever the same in strength of hand and foot, they took their pleasure in feasting, apart from all evils. When they died it was as though they were overcome by sleep. All good things were theirs and the grain harvest was yielded by bountiful earth of her own accord – abundantly, ungrudgingly – while they in peace and good will lived upon their lands with good things in abundance.[2]

Hesiod's Golden Age would have been dismissed as a fairy tale by someone like Thomas Hobbes, who perceived humanity's initial state in nature as a 'solitary, poor, nasty, brutish and short affair.' Today however, anthropologists would be more inclined to agree with Hesiod's interpretation of humankind's early history. Studies of the few remaining hunter-gatherer societies bear out much of Hesiod's account. Detailed examinations of the African bushmen and other hunter-gatherer groups provide some real surprises for those of us who like to believe that human history has been a progressive journey from the backbreaking toil and labour of the early primitives to the comfortable, leisurely life of twentieth-century America.

We moderns take pride in the fact that we only have to work 40 hours a week and that we can take off two or more weeks each year for vacation. Most hunter-gatherer societies would find such conditions intolerable. The fact is, contemporary hunter-gatherers work no more than twelve to twenty hours per week, and for weeks and months each year they do no work at all. Instead, their time is filled with leisure pursuits including games, sporting events, art, music, dance, ceremonies, and visiting with neighbours. Contrary to popular opinion, studies of the few remaining hunter-gatherer societies show that some are among the healthiest people in the world. Their diets are nutritious, and many – like the bushmen in Africa – live well into their sixties without the aid of modern medicine. Many hunter-gatherer societies place a premium on cooperation and sharing, and show little inclination for warring and aggression against each other or outside groups.

According to Hesiod, the Golden Age came to an abrupt end when Pandora lifted the lid on the box containing the evils of life. From then on each succeeding age has been more harsh and exacting than the one before it. The final age, according to Greek mythology, is the Iron Age. Here is Hesiod speaking in the eighth century before Christ:

For now in these latter days is the Race of Iron. Never by day shall they rest from travail and sorrow, and never by night from the hand of the spoiler. The father shall not be of one mind with the children, nor the children with the father, nor the guest with the host that receives them, nor friends with friends ... Parents shall swiftly age and swiftly be dishonoured ... The righteous man or the good man or he that keeps his oath shall not find favour, but they shall honour rather the doer of wrong and the proud man insolent. Right shall rest in might of hand and truth shall be no more.[3]

The Greeks believed that while the world was created by the Deity and was therefore perfect, it was not immortal. It had within it the seeds of decay. History, then, is the process whereby the original order of things maintains itself in perfection during the Golden Age, only to begin an inevitable process of decay during the subsequent ages of history. Finally, as the universe approaches ultimate chaos, the Deity intervenes once again and restores the original conditions of perfection. The whole process then begins once again. History is seen not as a cumulative progression toward perfection but as an ever repeating cycle moving from order to chaos.

This idea of history as a decaying cyclical process heavily influenced the Greek conception of how society should be ordered. Plato and Aristotle believed that the best social order was the one that experienced the fewest changes; there was no room in their world view for the concept of continued change and growth. Growth, after all, did not signal greater value and order in the world, but the exact opposite. If history represented the continued chipping away of the original perfect state, and the using up of the original fixed bounty, then the ideal state was the one that slowed down the process of decay as much as possible. The Greeks associated greater change and growth with greater decay and chaos. Their goal, then, was to hand down to the next generation a world as much preserved from 'change' as possible.

The Christian World View

Imagine, if you will, a time warp that could put you face to face with a medieval Christian serf. Now, the thirteenth century is not so very long ago. Only forty generations separate us from the feudal world. In fact, there's much about that world that you would immediately recognize. In England, students were already graduating from Cambridge, *Beowulf* had been written, and a form of English was being spoken – although we would find it difficult to understand. Still, even without a language barrier you and the serf would have very little of interest to say to each other after the usual chitchat about the weather. That's because you would probably be interested in finding out what his goals in life were. What contribution did he hope to make to the world? How was he bettering his lot in life? What kind of largesse did he expect to leave to his children? What were his ideas about happiness and the good life? You might even want to probe a little deeper into his psyche, asking him about his personality traits and identity problems.

Of course, you shouldn't expect much in the way of a response. In fact, if all you see in his eyes is a blank expression, it's not because you're talking over his head, or because his mind isn't developed enough for the exchange of ideas. It's just that his ideas about life, history, and reality are so utterly different from our own.

The Christian view of history, which dominated western Europe throughout the Middle Ages, perceived life in this world as a mere stopover in preparation for the next. The Christian world view abandoned the Greek concept of cycles but retained the notion of history as a decaying process. In Christian theology, history has a distinct beginning, middle, and end in the form of the Creation, the Redemption, and the Last Judgment. While human history is linear, not cyclical, it is not believed to be progressing toward some perfected state.

On the contrary, history is seen as an ongoing struggle in which forces of evil continue to sow chaos and disintegration in the earthly world.

Equally important, the doctrine of original sin precluded the possibility of humanity ever improving its lot in life. In fact, the idea of people making or changing history would have been unthinkable. After all, to the medieval mind, the world was a tightly ordered structure in which God controlled every single event. The Christian God was a personal God who intervened in every aspect of life. If things happened or didn't happen it was because God willed it. God made history, not people.

There were no personal goals, no desires to get ahead or to leave something behind. There were only God's decrees to be faithfully carried out. As historian John Randall points out, for the medieval Christian 'everything must possess significance not in and for itself, but for man's pilgrimage.' The purpose of every action, of every unfolding event, was tied to the 'purpose it served in the divine scheme.'[4]

The Christian world view provided a unified and all-encompassing picture of history. There was no room for the individual in this grand theological synthesis. It was duties and obligations, not freedoms and rights, that cemented and unified the historical frame of medieval life. Like the Greeks, the medieval concept of history was not one of growth and material gain. The human purpose was not to 'achieve things' but to seek salvation. Toward this end, society was viewed as an organic whole, a kind of divinely directed moral organism in which each person had a part to play.

Toward the Modern World View

There's no way to know how many professors have delivered lectures and how many students have been forced to sit through them over the course of history. Only a tiny handful of those lectures have ever made history. Jacques Turgot, a history teacher at the Sorbonne, is among those who have earned a place in this rather elite club. In 1750, he walked into a classroom in Paris, took out his notes, and began a two-part lecture in Latin on a new concept of world history. Turgot took on Plato, Aristotle, Saint Paul, Saint Augustine, and all the intellectual giants of the ancient and medieval worlds. By the time he had finished his last sentence and had tucked his lecture notes back into his valise, he had changed the entire construct of world history. The lectures, observed Frank Manuel, 'framed a new conception of world history from remotest antiquity to the present and constituted the first important version in modern times of the ideology of progress.'[5]

Turgot rejected both the cyclical nature of history and the concept of continued degradation. He argued, and rather pugnaciously, that history proceeds in a straight line and that each succeeding stage of history represents an advance over the preceding one. History, said Turgot, is both cumulative and progressive. Unlike the steady-state philosophers of Greece and the theologians of the Roman Church, he heralded the virtue of constant change and movement. Turgot was willing to acknowledge that progress is uneven and that occasionally it becomes bogged down or even retreats a few steps. Yet he held steadfast to the conviction that history demonstrates an overall advance toward the perfection of life here on earth. Bold thoughts! But Turgot's world was ready for them, as evidenced by the fact that Turgot was neither relieved of his appointment by the university nor condemned by the ecclesiastical authorities. A very radical change had taken place in the European mind between the time the giant cathedral at Chartres was being erected in the thirteenth century and that rather remarkable lecture in the year A.D. 1750. That change was the development of the modern world view. The story of its growth and maturation is the story of the world you and I have inherited.

Though we are largely unaware of it, much of the way we think, act, and feel can be traced back to the tiny strands and fragments that were woven together into the historical paradigm that took shape and form during those centuries of transition. It's ironic indeed that only now as that tapestry begins to fray and unwind is it possible to really see the stuff we and our modern world are made of.

The Machine Age

The modern age is the Machine Age. Precision, speed, and accuracy are the premier values. We are forever asking, 'How fast will it go?' or 'How long did it take you to get there?' The highest compliment we can pay is to say that something is so well made or thought out or planned that it virtually runs itself. We love the feel of metallic finishes, of aluminium, steel, and chrome. We find nothing more aesthetically pleasing than to ignite an engine or turn a switch. Our world is a world of pulleys and levers and wheels. Playtime is caught up with tinkering with contraptions; worktime with adjusting monitors and fine-tuning instruments. We regulate our daily routines by a machine – the watch. We communicate by a machine – the telephone. We learn by machines – the calculator, the computer, the television set. We travel by machines – the automobile, the jet. We even see by a machine – the electric light. The machine is our way of life and our world view rolled up in one. We view the universe as a grand machine set in motion aeons ago by the supreme technician, God. So perfectly engineered, it 'runs itself,' without ever missing a beat, and with a predictability of movement that can be calculated down to the nth degree.

We are mesmerized by the exactness we perceive in the universe, and we seek to duplicate its grandeur here on earth. History for us is a continuing exercise in engineering. The earth is like a giant hardware store, made up of all sorts of parts that need to be assembled together into a functioning system. Our job is never done. There are always new designs to consider and new jobs to be performed, all requiring the constant rearrangement of parts and the enlargement of processes. Progress then is 'geared' toward the perfection of the machine. There is the constant tying together of loose ends, the elimination of flaws, and the expansion of the machine process into every aspect of life. This is the historical paradigm of our age. We live by the dictates of the machine, and although we are quite willing to acknowledge its importance to our external way of life, we are much less willing to see how it has penetrated to the centre of our being.

The machine is now so firmly entrenched inside of us that it is difficult to know where it stops and we start. Even the words that come out of our mouths are no longer our words, they are the machine's words. We 'measure' relationships with other people by whether we are in 'synchronization' with them. Our feelings are reduced to good or bad 'vibrations.' We no longer initiate activity; instead we are a 'self-starter.' We avoid 'friction' at work and choose to 'tune in' rather than pay attention. We think of people's lives as either 'running smoothly' or 'breaking down.' If the latter, then we expect that in short order they will be put back together or 'readjusted.'

The Architects of the Mechanical World View

Every world view has its architects – those who sketch out the blueprint that the rest of us end up filling in. There were many preliminary drawings before the final plans for the Machine Age were agreed upon. By the middle of the eighteenth century all the key elements of the mechanical paradigm had been carefully integrated into a unified schema. The world was ready to turn the switch on the Machine Age. The mechanical world view is a testimonial to three men: Francis Bacon, René Descartes, and Isaac Newton. After 300 years we are still living off their ideas.

Francis Bacon laid the groundwork for the machine paradigm with a savage attack on the world view of the ancient Greeks. His *Novum Organum*, published in 1620, was a masterful piece of propaganda. Bacon sneered at the collected works of Plato, Aristotle, and Homer as nothing but 'contentious learning.'[6] The Greeks, he snapped, 'assuredly have that which is characteristic of boys; they are prompt to prattle but cannot generate; for their wisdom abounds in words but is barren of works.'[7] Bacon took stock of the Greek world view and concluded that, for all of its pompous claims, it had not 'adduced a single experiment which tends to relieve and benefit the condition of man.'[8] Bacon saw the world with different eyes. He didn't want to sit around contemplating nature. He wanted to find a methodology for controlling it. For the Greeks, the science of learning was intended to ask the metaphysical *why* of things; Bacon on the other hand thought that a science of learning should be committed to the *how* of things. 'Now the true and lawful goal of the sciences is none other than this: that human life be endowed with new discoveries and powers.'[9]

Some parts of Bacon's *Novum Organum* read more like an interoffice memorandum than a classical philosophical tract. For example, how many times have we heard our boss tell us to start dealing with the world as it is, not with how we would like it to be? Well, the boss is most likely unaware of it, but he's quoting Francis Bacon, who argued that we should begin 'building in the human understanding a true model of the world, such as it is in fact, not such as a man's own reason would have it to be.'[10] Bacon goes on to make it clear that a new method for dealing with the world is in order, one that can 'enlarge the bounds of human empire, to the effecting of all things possible.'[11] The new method Bacon alludes to is the scientific method, an approach that would separate the observer from the observed and provide a neutral forum for the development of 'objective knowledge.' According to Bacon, objective knowledge would allow people to take 'command over things natural – over bodies, medicine, mechanical powers and infinite others of this kind.'[12]

Bacon is the original no-nonsense pragmatist of the modern age. The next time you hear someone say to you, 'Try and be objective' or 'Prove it to me' or 'Just give me the facts,' think of Francis Bacon. He started it all off in 1620 with what he believed was a better idea for organizing the world.

Bacon had barely opened up the door to the new world view when René Descartes, a mathematician by trade, came barrelling through to announce the new floor plan. Shortly thereafter, he was followed in hot pursuit by Isaac Newton, who brought with him all the tools necessary to open up shop and begin doing business.

Descartes was not a modest man. He knew a good idea when he saw it. One rather cold day, according to his biographers, he was confined to his room by the severe weather. And that's when the idea hit him. The key to understanding the world, to deciphering its hidden secrets, to controlling it for human purposes was to be found in one word: *mathematics*.

As I considered the matter carefully, it gradually came to light that all those matters only are referred to mathematics in which order and measurement are investigated, and that it makes no difference whether it be in numbers, figures, stars, sounds or any other object that the question of measurement arises. I saw, consequently, that there must be some general science to explain that element as a

30

whole which gives rise to problems about order and measurement. This I perceived was called universal mathematics. Such a science should contain the primary rudiments of human reason, and its province ought to extend to the eliciting of true results in every subject.[13]

Descartes concluded with an observation that has since become the overriding axiom of the mechanics paradigm: 'To speak freely, I am convinced that it [mathematics] is a more powerful instrument of knowledge than any other that has been bequeathed to us by human agency, as being the source of all things!'[14] Here then was a man convinced, the first 'true believer' in the mechanical world view. Descartes wasted no time in popularizing his revelation. By the time he died in 1650, his mathematical view of nature had become accepted by the best minds all over Europe.

Descartes had succeeded in turning all of nature into simple matter in motion. He reduced all quality to quantity and then confidently proclaimed that only space and location mattered. 'Give me extension and motion,' he said, 'and I will construct the Universe.'[15] Descartes's mathematical world was tasteless, colourless, and odourless; it didn't ooze, drip, or spill. After all, what could be neater and more well behaved than algebra and geometry? Mathematics represented total order, and so in a single stroke of genius Descartes had successfully eliminated everything in the world which might in any way be thought of as messy, chaotic, and alive. In Descartes's world everything had its place and all relationships were harmonious. The world was one of precision, not confusion.

The Greek view of history as unfolding chaos and decay was deemed unmathematical and therefore false. The Christian world view fared little better. How could one ever know the workings of the natural order with precision if a personal God was constantly intervening in the affairs of life? In order to work as a world view, the mechanical paradigm had to be, above all, completely predictable. There was no room for a Divinity who could change the operating rules whenever He chose. God, then, was delicately retired from the scene. Of course, at first He was congratulated for being the supreme

31

mathematician who had engineered the whole plan and set it in motion, before going on to some other activity in the cosmic-theatre. Eventually, God was forgotten altogether, as succeeding generations became more and more intoxicated with the power this new-found paradigm provided them with.

Descartes gave human beings the 'faith' that they could unravel the truths of the world and become its masters. Newton then provided them with the tools they needed to do it. Newton discovered the mathematical method for describing mechanical motion. He argued that one law could explain why the planets move the way they do and why a single leaf falls from the tree in the manner it does. Subjecting all of nature to the laws of mathematics, Newton proclaimed that 'all the phenomena of nature may depend upon certain forces by which the particles of bodies, by some causes hitherto unknown, are either mutually impelled toward each other, and cohere in regular figures, or are repelled and recede from each other.' According to Newton's three laws, 'A body at rest remains at rest and a body in motion remains in uniform motion in a straight line unless acted upon by an external force; the acceleration of a body is directly proportioned to the applied force and is the direction of the straight line in which the force acts; and for every force there is an equal and opposite force in reaction.'[16] Soon after Newton published his mathematical method it was being taught at all the major universities. His fame spread to every corner of Europe, and when he died in 1727 he was given a royal funeral.

The mechanical world view dealt exclusively with material in motion, because that was the only thing that could be mathematically measured. It was a world view made for machines, not people. By separating and then eliminating all of the qualities of life from the quantities of which they are a part, the architects of the machine paradigm were left with a cold, inert universe made up entirely of dead matter. It was a short journey from the world as pure matter to the world of pure materialism, as we shall see in the next section.

Alfred North Whitehead delivered perhaps the single most devastating piece of commentary on the limitations of the Newtonian world machine as a historical paradigm. Noting that mechanics deals only with the space-time relationships of matter in motion, Whitehead remarked to his students:

As soon as you have settled . . . what you mean by a definite place in space-time, you can adequately state the relation of a particular material body to space-time by saying that it is just there, in that place: and, so far as simple location is concerned, there is nothing more to be said on the subject.[17]

The mechanical paradigm proved to be irresistible. It was simple, it was predictable, and above all it worked. Here, it appeared, was the long-sought-for explanation of how the universe functioned. There *was* an order to things, and that order could be ascertained by mathematical formulas and scientific observation. Still, as European scholars looked around them, they wondered why the normal activities of people in society often seemed so muddled and chaotic. The erratic behaviour of people and the imperfect workings of government and the economy didn't seem to square with the well-ordered mechanical explanation of the world that Bacon, Descartes, and Newton had put forth. The dilemma was quickly resolved: if society was misbehaving, then it could only be due to the fact that it was not adhering to the natural laws that govern the universe.

The only thing that was needed, then, was to figure out exactly how the natural laws applied to human beings and social institutions and then apply them. Obviously, this would be a long and difficult process – but no longer an impossible one, because the universal laws were now known. Besides, it would be well worth the time and effort, since the final payoff would be a perfectly ordered society. Humanity now had a new purpose in life. Gone was the medieval goal of seeking salvation in the next world. In its place was the new idea of seeking perfection in this world. History was now seen as a progressive journey from the rather disordered and confused state that society found itself in to the well-ordered and wholly predictable state represented by the Newtonian world machine.

Two men immediately set about the task of discovering the relationship between these universal laws and the workings of society. John Locke brought the workings of government and society in line with the world machine paradigm, and Adam Smith did the same with the economy.

Like most intellectuals of his period, Locke was deeply impressed with how the mechanical model had made sense out of a seemingly incomprehensible natural world. But why, he asked himself, were the affairs of human beings so chaotic? The answer, he concluded, was that the natural laws of society were being violated because the social order was built upon irrational traditions and customs that originated from the theocentrism that had ruled the world for so long. With the aid of reason, Locke set out to determine the 'natural' basis of society. He immediately concluded that religion could not form the social foundation simply because, by definition, God is unknowable. How can the unknowable be the proper basis for government? And so, in a monumental break with his philosophical predecessors, Locke argued that, while religion could rightly be a private concern of each person, it could not serve as the basis of public activity.

Having removed God from the affairs of people – as Bacon had removed Him from nature – Locke was left with human beings, all alone in the universe. No longer was the human being to be considered as part of a divinely directed organism. Now, men and women became just what Bacon, Descartes, and Newton had made of nature: mere physical phenomena interacting with other bits of matter in the cold, mechanical universe. This being the case, on what basis could a social order be formed? Here Locke provided an argument that has continued to dominate the modern world view down to the present. Once we cut through useless custom and superstition, argued Locke, we see that society, being made up solely of individuals creating their own meaning, has one purpose and one purpose only: to protect and allow for the increase of the property of its members. Pure self-interest thus becomes, in Locke's formulation, the sole basis for the establishment of the state. Society properly becomes materialistic and individualistic because, Locke maintains, reason leads us to conclude that this is the natural order of things. By the laws of nature, each individual is called upon to act out his role of

34

social atom, careering through life, attempting to amass personal wealth. There is no value judgment to be made here; self-interest is simply the only basis for society.

For Locke, the purpose of government was to allow people the freedom to use their new-found power over nature to produce wealth. Thus, from Locke's time to our own, the social role of the state has been to promote the subjugation of nature so that people might acquire the material prosperity necessary for fulfilment. 'The negation of nature,' Locke declared, 'is the way toward happiness.' People must become 'effectively emancipated from the bonds of nature.'[18]

But won't this constant and unmoderated scramble for personal affluence result in a savage war of each person against the other, with some members of society being victimized in the process? Not at all, says Locke, for human beings are not naturally evil or fallen, but inherently good. It is only scarcity and lack of property that make them evil. As people are naturally acquisitive, it is therefore only necessary to continue to increase the wealth of society and social harmony will continue to improve. People need not fight among themselves because nature has 'still enough and as good left; and more than the unprovided could use.'[19] People can have liberty of action because their self-interest would not conflict with others. Locke, then, became the philosopher of unlimited expansion and material abundance.

Still, are there no limits at all to the amount of wealth individuals can amass? After all, philosophers from Aristotle to Aquinas had argued that, beyond a certain point, property becomes a barrier to happiness. Not so, argues Locke. In a state of nature, he admits, it is true that primitives can only accumulate a limited amount of property from the bounty of nature. If a primitive attempts to take more property than his crude knowledge will allow him to consume, then it will spoil and possibly rob other members of the community of their own chance for accumulation. But in a commonwealth founded upon reason, where money as a medium of exchange exists, an unlimited amassing of property is permissible, indeed natural, for that is the purpose of money. Since money cannot possibly spoil, it is impossible to possess too much of it. Obviously, some individuals will amass more property than others, but this too is natural, for the world was given to 'the use of the industrious and rational.' He who applies reason the best will benefit the most.[20]

Locke does not stop here. The ownership of property (value extracted from nature) is not only a right in society; there is also a duty to generate wealth. In an environmentalist's nightmare, Locke writes that 'land that is left wholly to nature . . . is called, as indeed it is, waste.'[21] Nature is only of value when we mix our labour with it so that it will be productive:

He who appropriates land to himself by his labour, does not lessen but increases the common stock of mankind. For the provisions serving to the support of human life, produced by one acre of inclosed and cultivated land, are . . . ten times more than those which are yielded by an acre of land, of an equal richness lying waste in common. And therefore he that incloses land and has a greater plenty of the conveniences of life from ten acres than he could have from a hundred left to nature, may truly be said to give ninety acres to mankind.[22]

Using this early version of the 'trickle-down theory' (the more one individually makes, the more society collectively benefits), Locke goes on to declare that a person should 'heap up as much of these durable things (gold, silver, and so on) as he pleases; the exceeding of the bounds of his just property not lying in the largeness of his possession, but the perishing of anything uselessly in it.'[23] Reading Locke from our present-day concern with ecology, one has the unnerving feeling that he would not be satisfied until every river on earth were dammed, every natural wonder covered with billboards, and every mountain turned into rubble to produce oil shale. So rigidly productivist and materialistic is Locke that he condemns American Indians as a handful of people living in one of the richest lands in the world, idly refusing to exploit their riches: 'A king of a large and fruitful territory there feeds, lodges and is clad worse than a day-labourer in England.'[24]

With Locke, the fate of modern man and woman is sealed. From the time of the Enlightenment on, the individual is reduced to the hedonistic activities of production and consumption to find meaning and purpose. People's needs and aspirations, their dreams and desires, all become confined to the pursuit of material self-interest.

Like Locke, Adam Smith was enamoured of the mechanical world view and was determined to formulate a theory of economy that would reflect the universals of the Newtonian paradigm. In *The Wealth of Nations*, Smith argues that, just as heavenly bodies in motion conform to certain laws of nature, so too does economics. If these laws are obeyed, economic growth will result. But government regulation and control of the economy violated these immutable laws by directing economic activity in unnatural ways. Thus markets did not expand as rapidly as they could and production was stifled. In other words, any attempt by society to guide 'natural' economic forces was inefficient, and for Adam Smith, efficiency in all things was the watchword.

An inquiry into the laws of economics, Smith declared, will lead us to the inevitable conclusion that the most efficient method of economic organization is laissez-faire – the notion of leaving things alone and allowing people to act unhindered. Smith, like Locke, believed that the basis of all human activity is material self-interest. Since this is natural, we should not condemn selfishness by erecting social barriers to its pursuit. Rather, we should recognize people's desire to satisfy themselves for what it is – a virtuous activity that, in fact, benefits everyone. It is by each individual operating selfishly that scarcity may be overcome by surplus:

Every individual is continually exerting himself to find out the most advantageous employment for whatever capital he can command. It is his own advantage, indeed, and not that of society which he has in view. But the study of his own advantage naturally, or rather necessarily, leads him to prefer that employment which is most advantageous to the society.[25]

Smith explicitly removes any notion of morality from economics, just as Locke had done with social relations. Any attempt to impose morality on economy simply leads to a violation of the 'invisible hand,' which Smith asserted was a natural law that governs the economic process, automatically allocating capital investment, jobs, resources, and the production of goods. People could use reason to understand this law, Smith allowed, but just as human beings cannot control gravity, they cannot improve on the invisible hand. Since nothing can be more efficient than this 'natural' force controlling the rational market, wealth can best be produced only through free, unfettered trade and competition among rational, acquisitive individuals. Because the purpose of economics is a continually expanding market, anything that promotes growth is to be welcomed.

Believing that men and women are basically egoists in pursuit of economic gain, Smith's theories subordinate all human desires to the quest for material abundance to satisfy physical needs. There are no ethical choices to be made, only utilitarian judgments exercised by each individual pursuing self-interest.

Bacon, Descartes, Newton, Locke, and Smith were the great popularizers of the mechanical world view. Many others preceded and followed them. Still, their basic assumptions remain with us today. Those assumptions can be summarized in a few short sentences. First, there is a precise mathematical order to the universe that can be deduced from an examination of the motions of the heavenly bodies. Unfortunately, here on earth most things in the primal state are in a chaotic and confused condition. Therefore, things need to be rearranged to bring the same order to our world as appears to exist in the rest of the cosmos. The question then arises as to how best to arrange the stuff of nature so that it reflects the same kind of order that exists in the universe. The answer, it was assumed, was to use the scientific principles of mechanics to rearrange the stuff of nature in a way that best advanced the material self-interests of human beings. The logical conclusion to this grand new paradigm was simply this: *The more material well-being we amass, the more ordered the world must be getting*. Progress, then, is the amassing of greater and greater material abundance, which is assumed to result in an ever more ordered world. Science and technology are the tools for getting the job done. This, in a nutshell, is the chief operating assumption of the mechanical world paradigm.

The mechanical world paradigm has not been without its critics over the years. It has been ridiculed, attacked, and battered from many different quarters. Some of its assumptions have even been modified. Still, when one rereads Descartes, Locke, or Smith, one can't help but be impressed with how contemporary they sound. Every time a businessman, politician, or scientist speaks out in public on some pressing issue it's as if his speech had been ghostwritten by these long-dead seminal thinkers. Therefore, if the pronouncements tendered by our civic and public leaders seem more and more divorced from reality and less capable of explaining the problems facing our society, the blame isn't altogether theirs. If we're going to place the blame somewhere, then we should place at least part of it on Descartes, Locke, Smith, and their colleagues. After all, it's their methodology and ideas we're using.

The mechanical world paradigm experienced its greatest triumph in the aftermath of Charles Darwin's publication of *On the Origin of Species* in 1859. Darwin's theory of biological evolution was every bit as impressive as the scientific discoveries of Newton in physics. It could well have pushed the mechanical world view off centre stage and claimed hegemony for itself as a completely new organizing principle for society. It never happened. Instead Darwin's theories became an appendage to the Newtonian world machine. The full implications of Darwin's discoveries were never really explored. Instead, some of the more superficial trappings of his theory were immediately taken hold of and exploited in a way that further legitimized the mechanical world view.

Social philosophers like Herbert Spencer seized on Darwin's theory of the evolution of species as a kind of proof positive of the existence of progress in the world. Spencer and the so-called social Darwinists turned the concept of natural selection into the concept of the survival of the fittest. In so doing, they provided further support for the mechanical world view that holds that self-interest leads to increased material well-being, which leads to increased order.

Survival of the fittest was interpreted to mean that in the state of nature, each organism is engaged in a relentless battle with all other creatures. Those who survive and pass on their traits to their offspring are simply those best able to protect their own material self-interest. Evolution itself was seen as a process of ever-increasing order brought about as a result of each succeeding species' being better equipped to maximize its own self-interest and provide for its material needs. And so, Darwin's theory became a complete regurgitation of the chief assumption of the mechanical world view.

The mechanical age has been characterized by this notion of progress. Reduced to its simplest abstraction, progress is seen as the process by which the 'less ordered' natural world is harnessed by people to create a more ordered material environment. Or to put it another way, progress is creating greater value out of the natural world than what exists in its original state. Science, in this context, is the methodology by which people learn the ways of nature so that they can reduce them to consistent principles or rules. Technology, in turn, is the application of these rules in specific instances, the purpose being to transform parts of the natural process into workable forms of greater value, structure, and order than exist in the primal state.

The mechanical world view, the world view of mathematics, science, and technology, the world view of materialism and progress, the world view that claims to explain the world we experience, is beginning to lose its vitality because the energy environment upon which it was nourished is nearing its own death. (This argument will be examined later on in detail.) If there is a history to look back on, future generations will shake their heads in disbelief at the 400 years we call the modern age. The world as a machine will appear as naive to them as the Greeks' five ages of history have appeared to us. For they will be living under an entirely new world paradigm, one whose broad contours we will now explore.

PART TWO
The Entropy Law

The Entropy Law

An anthropologist, Max Gluckman, once remarked that 'a science is any discipline in which the fool of this generation can go beyond the point reached by the genius of the last generation.'[1] The first and second laws of thermodynamics are now taught in introductory physics courses. What they proclaim seems simple and commonsensical. Yet the path that led to their final articulation was an arduous one, littered with complex theories and the musings and speculations of many fine minds. Strangely enough, while scientists have anguished over the proper meaning of these two laws for longer than anyone cares to remember, they were already well established in the everyday folklore of just about every culture on earth. How many times have we heard the phrase 'You can't get something for nothing' or 'It does no good to cry over spilt milk' or 'You can't beat the system.' If you are familiar with these phrases and have seen them verified over and over again in your own everyday experience, then you know about the first and second laws of thermodynamics.

Thermodynamics sounds like a very complicated concept. In actuality it is both the simplest and at the same time the most impressive scientific conception we know of. Both laws of thermodynamics can be stated in one tiny sentence:

The total energy content of the universe is constant and the total entropy is continually increasing.[2]

What this means is that it is impossible to either create or destroy energy. The amount of energy in the universe has been fixed since the beginning of time and will remain fixed till the end of time. The first law is the conservation law. It says that while energy can never be created or destroyed it can be transformed from one form to another.

Science writer Isaac Asimov provides a simple example:

Suppose we take a quantity of heat and change it into work. In doing so, we haven't destroyed the heat, we have only transferred it to another place or perhaps changed it into another energy form.[3]

To be more specific, consider an automobile engine. The energy in the petrol is equal to 'the work done by the petrol motor, plus the heat generated, plus the energy in the exhaust products.'

The most important thing to remember, again, is that we cannot create energy. No person has ever succeeded in doing it and no person ever will. The only thing we can do is transform energy from one state to another. This is a heavy realization to come to when we stop to consider that everything is made out of energy. The shape, form, and movement of everything that exists is really only an embodiment of the various concentrations and transformations of energy. A human being, a skyscraper, an automobile, and a blade of grass all represent energy that has been transformed from one state to another. When a skyscraper or a blade of grass is formed, it is made of energy that has been gathered up from somewhere else. When the skyscraper is razed, and the blade of grass dies, the energy they embodied doesn't disappear. It is merely transferred back somewhere else into the environment. We've all heard it said that 'there's nothing new under the sun.' You can prove it to yourself with the next breath you take. You have just inhaled about 50 million molecules that were once inhaled by Plato.

If the first law of thermodynamics were all that we had to consider, then there would be no trick at all to using energy over and over again without ever running out of it. But we know that's not the way the world works. For example, if we burn a piece of coal, the energy remains but is transformed into sulphur dioxide and other gases that then spread out into space. While no energy has been lost in the process, we know that we can never reburn that piece of coal and get the same work out of it. The explanation for this is to be found in the second law of thermodynamics, which says that every time energy is transformed from one state to another 'a certain penalty is exacted.' That penalty is a loss in the amount of available energy to perform work of some kind in the future. There is a term for this; it's called entropy.

Entropy is a measure of the amount of energy no longer capable of conversion into work. The term was first coined by a German physicist, Rudolf Clausius, in 1868. But the principle involved was first recognized forty-one years earlier by a young French army officer, Sadi Carnot, who was trying to better understand why a steam engine works. He discovered that the engine did work because part of the system was very cold and the other part very hot. In other words, in order for energy to be turned into work, there must be a difference in energy concentration (i.e. difference in temperature) in different parts of a system. Work occurs when energy moves from a higher level of concentration to a lower level (or higher temperature to lower temperature). More important still, every time energy goes from one level to another, it means that less energy is available to perform work the next time around. For example, water going over a dam falls into a lake. As it falls, it can be used to generate electricity or turn a water wheel or perform some other useful function. Once it reaches the bottom, however, the water is no longer in a state to perform work. Water on a flat plane can't be used to turn even the smallest water wheel. These two states are referred to as *available or free energy states* versus *unavailable or bound energy states*.

An entropy increase, then, means a decrease in 'available' energy. Every time something occurs in the natural world, some amount of energy ends up being unavailable for future work. That unavailable energy is what pollution is all about. Many people think that pollution is a by-product of production. In fact, pollution is the sum total of all the available energy in the world that has been transformed into unavailable energy. Waste, then, is dissipated energy. Since according to the first law energy can neither be created nor destroyed but only transformed, and since according to the second law it can only be transformed one way – toward a dissipated state – pollution is just another name for entropy; that is, it represents a measure of the unavailable energy present in a system.

Now let's get back to Clausius, the man who thought up the word *entropy*. Clausius realized that in a closed system the difference in energy levels always tended to even out. Everyone who's ever taken a hot poker out of a fire has observed the same thing that Clausius made into a law. When a red-hot poker is removed from the fire and placed in the air, we soon notice that the poker begins to cool while the surrounding air begins to heat up. This is because the heat always flows from the hotter to the colder body. Finally, after enough time has elapsed, we can touch the poker and then place our hand in the surrounding air and, lo and behold, we find that they have reached the same temperature. The experts call this the equilibrium state, the state where there is no longer any difference in energy levels. This is the same state the water is in when on a flat plane. In both cases, the cooled-off poker and the flat water are no longer able to perform useful work. Their energy is bound energy or unavailable energy. Now that doesn't mean that the water can't be lugged up to the top of the dam again in buckets and dropped over or that the poker can't be reheated. But in each case, it means that a new source of free or available energy has to be used up in the process.

The equilibrium state is the state where entropy has reached a maximum, where there is no longer free energy available to perform additional work. Clausius summed up the second law of thermodynamics by concluding that 'in the world, entropy [the amount of unavailable energy] always tends toward a maximum.'

Here on earth there are two sources of available energy: our terrestrial stock and the solar flow from the sun. Economist Herman Daly explains the difference between the two:

The terrestrial stock consists of two kinds of resources: those renewable on a human time scale and those renewable only over geologic time and which, for human purposes, must be treated as nonrenewable. Terrestrial low-entropy stocks may also be classified into energy and material. Both sources, the terrestrial and the solar, are limited. Terrestrial nonrenewables are limited in total amount available. Terrestrial renewables are also limited in total amount available and, if exploited to exhaustion, become just like nonrenewables . . . the solar source is practically unlimited in total amount but strictly limited in its rate and pattern of arrival to earth.[4]

While the sun's energy is degrading with every passing second, its entropy will not reach a maximum until long after the earth's available terrestrial stock has been completely used up.

Every time you light a cigarette, the available energy in the world decreases. Of course, as already pointed out, it's possible to reverse the entropy process in an isolated time and place, but only by using up additional energy in the process and thus increasing the overall entropy of the environment. This should be especially understood when it comes to recycling. Many people believe that almost everything that we use up can be totally recycled and reused if only we develop the appropriate technology. This just isn't true. While more efficient recycling is going to be essential for the economic survival of the planet in the future, there is no way to achieve anywhere near 100 per cent reprocessing. For example, recycling efficiency today averages around 30 per cent for most used metals. Recycling requires the expenditure of additional energy in the collecting, transporting, and process-ing of used materials, which increases the overall entropy of the environment. Thus, things can only be recycled by the expenditure of new sources of available energy and at the expense of increasing the entropy of the overall environment.

A point that needs to be emphasized over and over again is that here on earth material entropy is continually increasing and must ultimately reach a maximum. That's because the earth is a closed system in relation to the universe; that is, it exchanges energy but not matter with its surroundings. With the exception of an occasional meteorite that falls to earth and some cosmic dust, our planet remains a closed subsystem of the universe. To those who mistakenly believe that the solar inflow of energy can be used to produce matter, economist Nicholas Georgescu-Roegen responds that 'even in the fan-tastic engine of the universe matter is not created from energy "alone" to any significant extent; instead, huge amounts of matter are continuously converted into energy.'[5] The point is, the sun, by itself, does not generate life. You can let the sun flow into an empty glass jar from now until the final heat death of the solar system and still no life will come forth. For life to unfold, the sun must interact with the closed system of matter, minerals and metals on the planet earth converting these materials to life and the utilities of life. This interaction facilitates the dissipation of this fixed endowment of terrestrial matter that makes up the earth's crust. Mountains are wearing down and topsoil is being blown away with each passing second. That is why, in the final analysis, even renewable resources are really nonrenewable over the long haul. While they continue to reproduce, the life and death of new organisms increase the entropy of the earth, meaning that less available matter exists for the unfolding of life in the future.

Every farmer understands that, even with recycling and constant sunshine, it's impossible to grow the same amount of grass on the same spot year after year in perpetuity. Every blade of grass grown today means one less blade of grass that can be grown sometime in the future on that same spot. That's because, like everything else, topsoil is part of the entropic flow. It contains the organic matter and inorganic minerals that allow the grass to grow. But the topsoil is only temporary. It begins as rock formations and organic refuse and much of it will end up as dust scattered into the wind or silt washed out to sea. In other words, topsoil is not a permanent fixture, but merely a particular concentration of matter along the entropic flow. In the short run (human time scale), it is possible to maintain the topsoil near a steady state as long as the erosion does not occur faster than nature can degrade rock formations and organic wastes into new topsoil. Even in the short run, however, topsoil often erodes faster than nature can replenish it, as a result of natural forces at work (wind storms, droughts, floods, etc.) or as a result of human intervention. Over-cultivating the land and the destruction of natural ecosystems often lead to de-mineralization of the soil and soil erosion, resulting in entropy patches for topsoil in isolated geographical pockets. It takes a thousand years to replace twelve inches of topsoil. Obviously, within the context of human time scales entropy of topsoil is a very real and continuous phenomenon. Matter is continually dissipating. The recognition of this fact was first advanced by

48

Nicholas Georgescu-Roegen: 'In a closed system, the material entropy must ultimately reach a maximum.'[6]

This is a difficult truth for most of us to accept because every child, when first introduced to elementary principles of biology, is taught that all matter recycles itself. Of course, this is true and is merely a restatement of the first law of thermodynamics, that matter (and energy) can neither be created nor destroyed. Unfortunately, the second law of thermodynamics is generally ignored. It tells us that while matter is continually recycled, a price has to be paid each time in terms of degradation. For example, suppose we extract a chunk of metallic ore from beneath the earth's surface and fashion it into a utensil. During the life time of that utensil, metal molecules are constantly flying off of the product as a result of friction, and wear and tear. Those loose metal molecules are never destroyed. They eventually find their way back into the earth. But now they are randomly dispersed throughout the soil and are no longer in a concentrated form to perform useful work, like the original chunk of metallic ore. A way might be found to recycle all of these randomly dispersed metal molecules but only at the expense of an increase in entropy in the process. A mechanical device would have to be assembled to recollect the metal molecules and an energy source introduced to run the machine. Since the machine itself is made out of metallic ore from the earth, it would be losing its own metal molecules to friction and wear and tear even as it is recycling the other random metal molecules. At the same time, the energy used to run the recycling machine would also end up increasing the entropy.

When energy becomes unavailable we use the term heat death. When matter becomes unavailable we use the term 'matter chaos.' The result in both cases is entropy; a randomization of matter and energy making both less concentrated and thus less fit to perform useful work.

Some scientists have argued that in the very long run, the sun, acting upon the earth's crust, might somehow reconstitute all of the random metal molecules into a concentrated state once again. This may be statistically possible, but of little help to the human species since the time frame being discussed is measured in geological units, namely billions of years. In the short run and in specific geographical pockets, entropy of matter and energy is a very real observable phenomenon.

The Entropy Law is something that needs to be felt as much as understood. The essence of this law is the essence of reality itself, and so getting hold of its meaning requires a kind of intuition. For this reason, it will be helpful to look at the Entropy Law from some other directions.

Another way of talking about energy levels and entropy has already been touched upon – namely concentrations. Why is it that when you open up a bottle of perfume, the odour begins to escape into the air, and after a short time it pervades the room? Or let's say we open the door of the room into an even larger room only to find, a few minutes later, that the perfume can now be smelled in both rooms although the smell is much less intense than when it was in only one room. Bertrand Russell explains the process:

Whenever there is a great deal of energy in one region and very little in a neighbouring region, energy tends to travel from the one region to the other, until equality is established. This whole process may be described as a tendency towards democracy.[7]

Again, this is just another way to understand the second law. Energy always moves from the more concentrated state (in this case the perfume bottle) to the less concentrated state (both large rooms). In the process, free or available energy is used up or dissipated (the odour loses its potency). If you were to look at the perfume on a molecular level, you would notice that while they are cooped up in the bottle, the molecules are bombarding each other at an incredibly fast rate. As soon as they are allowed to escape from the bottle, however, the molecules begin their random journey into the larger space. As they begin to spread out over the room they collide with each other less frequently until they are uniformly distributed throughout the whole room.

There have been many attempts to find a way around the Entropy Law. In fact, it's been one of the favourite pastimes of scientists and philosophers alike. Perhaps the most impressive challenges to the Entropy Law came from two highly respected scientists, J. C. Maxwell and Ludwig Boltzmann, in the late nineteenth century. Both challenges only ended up strengthening the position of the second law, and for that reason they deserve mentioning.

Maxwell suggested that an intelligent being tiny enough to handle individual molecules might be capable of violating the second law. No matter that we have yet to come across any such little fellow; the argument is still interesting for what it says about the lengths the scientific community was willing to go to try to overcome the second law.

Maxwell posed the following hypothesis. Take an enclosure, he said, that is divided into two compartments, separated by a small door. The enclosure, which is totally isolated, contains a gas at a 'uniform temperature.' Now at uniform temperature the Entropy Law says that no work can be performed. Maxwell proposed to get around that problem by putting a little demon at the tiny door separating the two compartments. The demon, being sharp of eye, would then open and close the door, permitting molecules with greater than average velocities to pass from left to right and molecules with less than average velocities to pass from right to left. 'Since high speed molecules correspond to a high temperature and low speed molecules to a low temperature, the gas in the right-hand compartment would become hotter and the gas in the left-hand compartment colder.' Need we say more? 'Once the difference in temperature was established, it could be used to drive a heat engine that would deliver useful work.'[8]

Starting from maximum entropy or a total equilibrium state of uniform energy, Maxwell proposed to reverse the entropy process without any outside energy being used; this would have violated the second law. First, it's obvious that in the real world we'd never be able to produce such a demon. But just to humour Maxwell, let's assume an appropriate demon could be found and that it would be willing to take on the job. Could it perform its work without violating the second law? Stanley Angrist and Loren Hepler, writing in *Texas Quarterly*, put the demon to the test and discovered that even it could not get around the Entropy Law:

[Maxwell] supposed that his demon would be able to sense the velocity (speed and direction) of individual molecules and then act accordingly ... As the demon peers into either side of the isolated enclosure at uniform temperature, the uniformity of radiation throughout does not permit him to see anything. The sameness in the enclosure would allow him to perceive the thermal radiation and its fluctuations, but he would never see the molecules ... We conclude that the demon needs his own supply of light to disturb the radiation equilibrium within the enclosure, so we equip him with a light to enable him to see the molecules. The high quality energy that the light pours into the system provides the demon with the information he needs to operate the door to separate the high speed molecules from the low speed ones. Although the demon is able to increase the net order of the gas (and hence decrease its total entropy), a greater increase in disorder and entropy must occur in the light source. That is, for the entire system, light source, demon and gas, there will be a net increase in entropy as required by the second law, thus rendering the perpetual motion machine impossible.[9]

About the only thing this whole exercise proves is that 'we cannot get anything for nothing, not even an observation.'[10]

Maxwell's attempt to challenge the Entropy Law is worth remembering. It is, more than anything else, a reflection of the hardheaded refusal of the scientific community to acknowledge the full implications of what the Entropy Law means for science, philosophy, and life on this planet.

Adding embarrassment to fantasy, Ludwig Boltzmann jumped into the fray, determined to rescue classical physics from the steady encroachment of the Entropy Law. Boltzmann's 'h-Theorem' is a remarkable sleight of hand designed to accommodate the second law while at the same time undermining its clout. Boltzmann acknowledged the validity of the second law up to a point. He was willing to admit that in a closed system, entropy increases, but was unwilling to claim that it was an absolute certainty. He preferred the word *probably* to *certainly* and in so doing attempted to turn the second law into a probability or statistical law. What Boltzmann was saying is that while it's unlikely that energy would move from a colder to a hotter state, it was not impossible. It's important to be clear on what Boltzmann was arguing because it is still taken seriously by many scientists. Sir Arthur

Eddington gets right to the point about the likelihood of Boltzmann's probability theorem ever working, even once, in the real world. He proposes a vessel with two equal parts separated by a partition. The first compartment contains air, the second compartment a vacuum. The partition between the two compartments is opened, allowing the air to spread evenly through the vessel. Eddington allows that at some future time there is always the chance that all of those billions upon billions of molecules of air diffused through the entire vessel will in their individual random movements all end up in the right-hand side of the compartment once again at exactly the same time. As to how probable such an occurrence is, Eddington concludes:

If an army of monkeys were strumming on typewriters they 'might' write all the books in the British Museum. The chance of their doing so is decidedly more favourable than the chance of the molecules returning to one half of the vessel.[11]

Even more to the point is Nicholas Georgescu-Roegen. He is worth quoting at length because his criticism of statistical thermodynamics zeroes in on the battle between the mechanical paradigm and the emerging entropy paradigm.

It must be admitted, though, that the layman is misled into believing in entropy bootlegging by what physicists preach through the new science known as statistical mechanics but more adequately described as statistical thermodynamics. The very existence of this discipline is a reflection of the fact that, in spite of all evidence, man's mind still clings with the tenacity of blind despair to the idea of an actuality consisting of locomotion and nothing else. A symptom of this idiosyncrasy was Ludwig Boltzmann's tragic struggle to sell a thermodynamic science based on a hybrid foundation in which the rigidity of mechanical laws is interwoven with the uncertainty specific to the notion of probability ... According to this new discipline, a pile of ashes may very well become capable of heating the boiler. Also, a corpse may resuscitate to lead a second life in exactly the reverse order of the first. Only, the probabilities of such events are fantastically small. If we have not yet witnessed such 'miracles,' the advocates of statistical mechanics contend, it is only because we have not been watching a sufficiently large number of piles of ashes or corpses.[12]

We have looked at the second law from the perspective of energy moving from available to unavailable states and from high concentrations to low. There is still another way to view the second law, the most profound way of all. The Entropy Law is also a statement that all energy in an isolated system moves from an ordered to a disordered state. The minimum entropy state, where concentration is highest and where available energy is at a maximum, is also the most ordered state. In contrast, the maximum entropy state, where available energy has been totally dissipated and diffused, is also the most disordered state.

This conforms to our everyday sense of the world around us. Left on their own, things do not tend to spontaneously move to more and more ordered states. Anyone who has ever had to take care of a house, or work in an office, knows that if things are left unattended they soon become more and more disorderly. Bringing things back into a state of order requires the expenditure of additional energy. For example, consider a deck of playing cards that is organized by number and suit. The deck is in a state of maximum order or minimum entropy. Fling the deck to the ground and the cards will scatter into a random, disordered state. Picking each card off the floor and then arranging them one by one back into their original ordered state will take the expenditure of more energy than was used to scatter them in the first place.

It must be emphasized that whenever the entropy increase is reversed in one place, it is only done by increasing the overall entropy of the surrounding environment. This is so because every time an event occurs, some amount of energy is dissipated in the process and thus made totally unavailable for future use. This dissipated energy is added to the sink of dissipated energy that has accumulated as a result of the occurrence of every other past event. The tremendous implications for society that flow from this are truly mind-boggling. To quote Angrist and Hepler: 'Each localized, man-made or machine-made entropy decrease is accompanied by a greater increase in entropy of the surroundings, thereby maintaining the required increase in total entropy.'[13]

Albert Einstein once mused over which of the laws of

science deserved to be ranked as the supreme law. He concluded by making the following observation:

A theory is more impressive the greater is the simplicity of its premises, the more different are the kinds of things it relates and the more extended its range of applicability. Therefore, the deep impression which classical thermodynamics made on me. It is the only physical theory of universal content which I am convinced, that within the framework of applicability of its basic concepts will never be overthrown.[14]

Cosmology and the Second Law

Whenever scientists begin speculating about the second law, the question ultimately arises as to how broadly it can be applied. For example, does the Entropy Law apply to the macroworld of stars and galaxies that make up the universe? In fact, the Entropy Law is the basis of most cosmological theories. Scientist Benjamin Thompson became the first to draw the cosmological implications of the second law back in 1854. According to Thompson, the Entropy Law tells us that

within a finite period of time past, the earth must have been, and within a finite period of time to come the earth must again be, unfit for the habitation of man as at present constituted, unless operations have been, or are to be performed, which are impossible under the laws to which the known operations going on at present in the material world are subject.[15]

Two years later Helmholtz formulated what has become the standard cosmological theory based on the Entropy Law. His theory of 'heat death' stated that the universe is gradually running down and eventually will reach the point of maximum entropy or heat death where all available energy will have been expended and no more activity will occur. The heat death of the universe corresponds to a state of eternal rest.

Today the most widely accepted theory about the origin and development of the universe is the big bang theory. First conceptualized by Canon Georges Lemaître, the big bang theory postulates that the universe began with the explosion of a tremendously dense energy source. As this dense energy expanded outward, it began to slow down, forming galaxies, stars, and planets. As the energy continues to expand and become more diffused, it loses more and more of its order and will eventually reach a point of maximum entropy, or the final equilibrium state of heat death. The big bang theory coincides with the first and second laws. It states that the universe started with complete order and has been moving toward a more and more disordered state ever since. If this theory appears familiar, it should. Both the ancient Greek and the medieval Christian view of history share much in common with the cosmologists' notion of the history of the universe.

It's strange indeed that we in the modern world are willing to see the history of the universe as beginning with a perfect state and moving toward decay and chaos and yet continue to cling to the notion that earthly history follows the exact opposite course, i.e., that it is moving from a state of chaos to a 'progressively' more ordered world. This contradiction is so blatant that it should not come as too big a surprise that there have been attempts to formulate other cosmological theories to get around the Entropy Law. For a few years the 'theory of continual creation' was the vogue. Back in 1948 three young scientists, Fred Hoyle, Thomas Gold, and Herman Bondi, suggested that while the universe is definitely expanding, heat death or maximum entropy could be avoided by interjecting negative entropy into the universe from 'without.' If just the right amount of new negative entropy were introduced to compensate for entropy loss, then the universe would continue to go on forever with new galaxies forming at the same time as others are burning out, like a kind of cosmic perpetual motion machine. So, while there would be losses in parts of the universe, the gains in other parts of the cosmic theatre would assure that the entire system never degraded. Unfortunately for Hoyle, Gold, and Bondi, subsequent scientific experiments have invalidated their theory. In the 1960s, astronomers began counting the number of radio sources back through time and out into space. In order for the continual creation theory to be proven correct they would at least need to show that there had been no substantial change in radio sources between the past and present. The results of the experiment proved devastating to the continual creation theory, because they showed that there were more radio sources in the distant past than there are now, thus reconfirming the big bang theory and the second law, that the entropy of the universe is moving toward a maximum and heat death.

Other evidence continued to come in invalidating the continual creation theory and offering additional support for the big bang theory. It was found that quasars, which are some of the most distant objects known, were, like radio sources, much more numerous in the past. Finally two scientists, Penzlas and Wilson, delivered the crushing blow to the continual creation theory with their discovery of 'universal background heat radiation.' There was simply no way to account for this phenomenon in the continual creation theory of the operation of the universe.

There have been other theories as well. The cyclical theory, for example, holds that the universe is forever moving through endless series of expanding and contracting phases without beginning or end. According to this theory, the last big bang is just one of an infinite chain of big bangs which have occurred and will continue to occur forever. As the present expanding universe reaches maximum entropy, it will then begin to contract back to a more and more ordered state until the entire universe is condensed into a critical mass the size of an atomic nucleus, at which time it will explode back into the cosmic reaches once again. At the present time, the cyclical theory remains highly speculative, since so little experimentation has been done to confirm or refute its central thesis. For the moment all we can say for sure is that, for our tiny solar system and the planet earth, the Entropy Law still holds 'the supreme position among the laws of nature.'

Time, Metaphysics, and Entropy

Nowhere is the Entropy Law more important than in the determination of time. Saint Augustine once wrote, 'I know what time is, if no one asks me, but if I try to explain it to one who asks me, I no longer know.'[16] The mechanical world view of time is very different from the entropy world view. In classical physics, time can go in either direction. Because Newtonian principles are based on mathematics, every change in matter in motion must be reversible in theory. For example, imagine a film showing billiard balls colliding with each other. Now let's reverse the film and run it backwards. It still seems to make perfect sense, even in the reverse order. As long as we are dealing with simple matter in motion, in the Newtonian sense, time can be represented equally well as both $+T$ and $-T$. But now suppose we run a second film showing water plunging over Niagara Falls. As soon as we reverse this second film, everything looks ridiculous. The water is now flowing from the bottom up to the top of the falls. While the Newtonian model based on mathematics tells us that, in theory, the water could reverse itself and run uphill, we know it can't happen. The reason is explained by the second law.

'Time waits for no one.' 'Time goes on.' 'You can't go back in time.' Indeed! The point is, time as we experience it is irreversible. Time only goes in one direction, and that's forward. That forward direction, in turn, is a function of the change in entropy. Time reflects the change in energy from concentration to diffusion or from order to increasing disorder. If the entropy process could be reversed, then everything that has been done could be undone. In the words of Lord Kelvin: 'Boulders would recover from the mud and would become reunited to the mountain peak from which they had formerly broken away.'[17]

Time goes forward because energy itself is always moving

from an available to an unavailable state. Our consciousness is continually recording the entropy change in the world around us. We watch our friends get old and die. We sit next to a fire and watch its red-hot embers turn slowly into cold white ashes. We experience the world always changing around us, and that experience is the unfolding of the second law. It is the irreversible process of dissipation of energy in the world. What does it mean to say, 'The world is running out of time'? Simply this: we experience the passage of time by the succession of one event unfolding after another. And every time an event occurs anywhere in this world energy is expended and the overall entropy has been increased. To say the world is running out of time, then, is to say the world is running out of usable energy. In the words of Sir Arthur Eddington, 'Entropy is time's arrow.'

Both the ancient Greek and the medieval Christian world views, with their idea of history as a process of movement from order to decay, reflected an understanding of the true direction of time's arrow and the entropy process. By ignoring the truths of the Entropy Law, the existing world paradigm of Newtonian mechanics has provided the illusion that time is an autonomous process in the world, independent of the workings of nature. This sense of alienation from nature began with Descartes's suggestion that the world is organized in such a way that there is total separation between people and nature. The crux of the scientific method is the establishment of complete neutrality between the observer and the observed, so that nature could be manipulated and used to advance the material interest of humankind.

Having hit upon a method of organizing the world that effectively separated people from nature, the true relationship between life, time, and the entropy process was severed from people's consciousness. From there it's not hard to understand how Locke and his friends could devise a world view that went completely counter to the real workings of the world. While the Entropy Law states that all things in nature can only be transformed from a usable to an unusable state, Locke argued the opposite. Claiming that everything in nature was waste until people took hold of it and transformed

60

it into usable things of value, he and the other architects of the existing mechanical paradigm argued that the world was in fact 'progressing' from chaos to order. As to the passage of time, they reasoned that the faster nature was transformed, the more progress would occur, the more ordered the world would become, and the more time would be saved.

This view of time and history is completely backwards. As already mentioned, time can only exist as long as there is available energy to perform work. The amount of real time expended is a direct reflection of the amount of energy used up. As the universe runs out of available energy, fewer and fewer occurrences can happen – which means less and less 'real' time is still available. Eventually, as the final equilibrium state of heat death is reached, everything will stop occurring. Time, then, will no longer exist as we experience it, because nothing will any longer be occurring. Therefore, the faster the energy of the world is used up, the fewer are the possible occurrences left that can unfold, and, correspondingly, the less time that's left in the world. The fact is, we never save time by expending greater amounts of energy. On the contrary, the more energy we expend, the more time we use up. The next time someone asks you how much time you saved by expending more energy to do a particular thing faster, think about the Entropy Law and time's arrow, and then think about the peculiar way we have come to view history over the past 400 years.

There is one more aspect of entropy and time that deserves attention. While entropy tells us the direction of time, it doesn't tell us the speed. The fact is, the entropy process is constantly changing speed. With every occurrence in the world, entropy increases – but sometimes slower, sometimes faster. Its speed depends upon how many babies are being born, how many blades of grass are dying, how many cars are being built, how many raindrops are falling to the ground, how much wind is blowing, and how many pebbles are being ground into sand as the waves wash up on the beaches of the world.

Humans have always debated the question of whether history is predetermined, or whether we are able to exercise a measure of free will over the unfolding of events. The Entropy Law, more than any other concept we have discovered, goes a long way toward resolving that question. In establishing the direction of time, the second law sets the limits we are forced to work within. We cannot reverse time or the entropy process. It is determined for us. But we can exercise free will in determining the speed by which the entropy process moves. Every action we human beings take in this world either speeds up or slows down the entropy process. By the way we choose to live and behave, we decide how quickly or how slowly the available energy in the world is dissipated. Here is the point where science joins metaphasics and ethics. The full implications of this juncture between free will, determinism, and the entropy process will be explored in depth in later sections when we look at the nature of technology and economic theory.

Life and the Second Law

If the overall entropy of the world is always increasing, then how do we explain the process of life? Certainly, living things exhibit a great deal of order. Evolution itself appears to represent the continued accumulation of greater and greater order from disorder. No one would deny that as a little baby continues to grow it stores up greater amounts of energy. Every time we look at a plant or animal we marvel at how well organized are all the billions of molecules that make it up. Life, then, must violate the second law, right? Wrong! For a long time, scientists were confused on this point. Now they acknowledge that, like everything else in the world, life cannot escape the iron hand of the Entropy Law. Says Harold Blum, in his pioneering book on the subject, *Time's Arrow and Evolution*, 'The small local decrease in entropy represented in the building of the organism is coupled with a much larger increase in the entropy of the universe.'[18]

Living things are able to move in a direction opposite to that of the entropy process by absorbing free energy from the surrounding environment. The ultimate source of that free energy is the sun. All plant and animal life is dependent on the sun for survival – either directly, in the case of plants performing photosynthesis, or indirectly, in the case of animals that eat plants or other animals. Every living thing survives, in the words of the Nobel Prize-winning physicist Erwin Schrodinger, 'by continually drawing from its environment negative entropy ... What an organism feeds upon is negative entropy; it continues to suck orderliness from its environment.'[19]

In other words, in all living things the natural tendency is to move toward equilibrium. We human beings, for example, are constantly dissipating our energy every time we think a thought or twitch a finger. In order to prevent ourselves from dissipating to an equilibrium state of death, we require a constant flow-through of free energy (negative entropy) from our larger environment. Anyone not convinced of this truth has probably never seen a dead body. Within hours of death, the body begins to completely unravel, dissipating into total random chaos.

The reason scientists had such a hard time figuring out how living systems fit into the second law is because equilibrium thermodynamics is concerned with closed systems – systems in which energy but not matter can be exchanged with the outside surroundings. Living systems, however, are open systems. Both matter and energy are exchanged with the outside. Living systems can never obtain an equilibrium state, while they're alive, because an equilibrium state means death. So, living things maintain themselves far away from an equilibrium state by continuing to feed off the available energy around them. This state is called the 'steady state.' If matter and energy cease flowing through a living organism, the steady state is abandoned, and the organism drifts to equilibrium and death. In living systems then, free energy flow, not entropy, is the primary concern. This branch of science is called nonequilibrium thermodynamics. While nonequilibrium systems cannot be explained in the same way as equilibrium systems, they do conform with the broad imperative laid down by the second law, as we shall continue to see.

'Every living thing,' said Bertrand Russell, 'is a sort of imperialist, seeking to transform as much as possible of its environment into itself and its seed.'[20] In this process of energy scavenging, every living thing on this planet dissipates energy as that energy flows through its system, making at least part of it unavailable for future use. It is also true that even the tiniest plant maintains its own order at the expense of creating greater disorder in the overall environment. In the case of the plant, it survives by photosynthesis – sucking negative entropy from the sun's rays. In the process, only a tiny fraction of the solar energy is actually picked up and used by the plant; the rest is simply dissipated. Compared with the tiny entropy decrease in the plant, the energy lost to the overall environment is monumental.

The entropy increase is even more graphically illustrated in the normal food chain. Chemist G. Tyler Miller sets up a very simple food chain to make the point. The chain consists of grass, grasshoppers, frogs, trout and humans. Now, according to the first law, energy is never lost. But according to the second law, available energy should be turned into unavailable energy at each step of the food chain process, and therefore the overall environment should experience greater disorder. In fact, this is exactly what happens. At each stage of the process, when the grasshopper eats the grass, and the frog eats the grasshopper, and the trout eats the frog, and so on, there is a loss of energy. In the process of devouring the prey, says Miller, 'about 80–90% of the energy is simply wasted and lost as heat to the environment.'[21] Only between 10 and 20 per cent of the energy that was devoured remains within the tissue of the predator for transfer to the next stage of the food chain. Consider for a moment the numbers of each species that are required to keep the next higher species from slipping toward maximum entropy. 'Three hundred trout are required to support one man for a year. The trout, in turn, must consume 90,000 frogs, that must consume 27 million grasshoppers that live off of 1000 tons of grass.'[22]

Thus, in order for one human being to maintain a high level of 'orderliness,' the energy contained in 27 million grasshoppers or a thousand tons of grass must be used. Is there any doubt, then, that every living thing maintains its own order only at the expense of creating greater disorder (or dissipation of energy) in the overall environment?

Energy is continuously flowing through every living organism, entering the system at a high level and leaving the system in a more degraded state. Organisms survive by being able to accumulate negative entropy from their environment. The struggle for existence depends upon how well equipped each organism is to capture available energy. Biologist Alfred Lotka was one of the first to relate energy flow-through and biological evolution. Lotka said that every species can be looked at as a different type of 'transformer' for capturing and using available energy. Each transformer or organism is equipped with an array of devices that it uses to suck in energy from its surroundings.

According to Lotka, 'The close association of the principal sense organs: eyes, ears, nose, tastebuds, tactile papillas of the finger tips, with the anterior [head] end of the body, the mouth end, all point to the same lesson.'[23] That lesson is that organisms are designed to be collectors and transformers of energy. If they weren't so designed, they wouldn't be able to survive. From the point of view of evolution, Lotka argues that natural selection favours those organisms that are able to 'increase the total mass of the system, rate of circulation of mass through the system, and the total energy flux through the system ... so long as there is presented an unutilized residue of matter and available energy.'[24]

Lotka's assertion that natural selection favours those organisms that maximize the flow of energy through the system has since been modified (even by Lotka himself). It is now acknowledged that maximizing flow-through is a common response in the early stages of an ecological system's development, when there is still an excess of available energy present. However, as various species begin to fill up a given ecological habitat, they are forced to adapt to the ultimate carrying capacity of the environment by using less energy flow-through more efficiently. The early stage of maximum flow-through is generally referred to as the colonizing phase, and the later stage of minimum flow-through as the climactic phase.

On the whole, Homo sapiens has yet to move from a colonizing to a climactic phase. Human beings, especially in the highly industrial societies, continue to order activity in such a way as to increase energy flow-through in both the human and the social systems. The worldwide human crisis today is a crisis of transition. In the next age, humanity will have settled into its climactic phase, ordering its activity in such a way as to minimize energy flow-through in the human and social processes. If it doesn't, it will likely go the way of other species who were unable to make the transition in the past. Life's epic is strewn with extinct species; it would have little trouble accommodating at least one more on the long list of names.

We are so used to thinking of biological evolution in terms of progress. Now we find that each higher species in the evolutionary chain transforms greater amounts of energy from a usable to an unusable state. In the process of evolution, each succeeding species is more complex and thus better equipped as a transformer of available energy. What is really difficult to accept, however, is the realization that the higher the species in the chain, the greater the energy flow-through and the greater the disorder created in the overall environment.

The Entropy Law says that evolution dissipates the overall available energy for life on this planet. Our concept of evolution is the exact opposite. We believe that evolution somehow magically creates greater overall value and order on earth. Now that the environment we live in is becoming so dissipated and disordered that it is apparent to the naked eye, we are for the first time beginning to have second thoughts about our views on evolution, progress, and the creation of things of material value. More about the implications of this in later sections.

Explanations and rationalizations aside, there is no way to get around it. Evolution means the creation of larger and larger islands of order at the expense of ever greater seas of disorder in the world. There is not a single biologist or physicist who can deny this central truth. Yet, who is willing to stand up in a classroom or before a public forum and admit it?

If you feel like you've just been hit on the head with a sledgehammer and that this whole explanation of evolution is simply too depressing to consider, it's only because we are so locked into the existing world paradigm that all other ways of organizing our thoughts seem totally unacceptable. Yet, not until we recognize and acknowledge that the second law is the real basis of both life and evolution will we be able to make the transition from our present colonizing phase to a climactic phase of existence.

Exosomatic Instruments and Energy

While all living things are engaged in a continuous struggle to suck available energy from their surroundings, only Homo sapiens is equipped with external aids to help facilitate the process. Other creatures must rely on their own anatomy – their eyes, ears, nose, teeth, claws, and so on – to gather up energy. Human beings, however, because of our more highly developed nervous system and brain, have succeeded in augmenting and extending our natural biological apparatus with the creation of all sorts of tools. Scientists and anthropologists refer to these instruments as exosomatic in nature to distinguish them from the endosomatic organs we're born with.

When we talk about exosomatic instruments, we're really including the entire range of tools that people use to capture, transform, and process available energy (or negative entropy) through our systems. We develop tools and machinery to extract energy from the environment; we build homes to capture heat and maintain our body warmth; we build roads, construct bridges, and engineer new ways of travelling to facilitate the transport of energy from one location to another; we devise languages, customs, economic institutions, and governments to better organize the processing and distribution of energy.

All of these exosomatic activities together constitute a large part of human culture. Social development, after all, is basically the attempt to create pockets of order to advance human survival. Since we human beings, like all other living creatures, survive by our ability to maintain a constant flow-through of energy, our cultures serve as an instrument for the withdrawal of energy from the larger environment. The first and second laws of thermodynamics, then, serve as the supreme operating principles for every culture and civilization, just as they do for the rest of the universe. The belief that we can leave these laws outside the city gates is as dangerous as the notion that we can survive without a constant flow of energy through our systems.

If we were to abstract all of the complex activities that go on in a culture into a few categories, the terms *transforming*, *exchanging*, and *discarding* of energy would no doubt top the list. People are always busily engaged in one or more of these processes. Yet, it's often hard for us to see these processes for what they are, because they have become so associated with activities that seem to have no relationship whatsoever to nature itself. Still, if we take the time to strip away all of the layers upon layers of cultural paint that have accumulated over the centuries, we will notice that at the bottom, it is really available energy that is being constantly transformed, exchanged, and discarded. If this sounds hard to believe, test it out yourself. Take an entire day to observe everything you come in contact with: things that you see, hear, touch, smell, feel, or consume; things that you change; and things that you exchange. Then try to trace each experience or item in both directions, back to its original source and forward to its final destination. Chances are better than excellent (in fact, guaranteed) that they all started off as some form of raw material (available energy), and that they will all end up somewhere as unusable waste (unavailable energy).

Energy is the basis of human culture, just as it is the basis of life. Therefore, power in every society ultimately belongs to whoever controls the exosomatic instruments that are used to transform, exchange, and discard energy. Class division, exploitation, privilege, and poverty are all determined by how a society's energy flow line is set up. Those who control the exosomatic instruments control the energy flow line. They determine how the work in society will be divided up and how the economic rewards will be allocated among various groups and constituencies.

It's a bizarre experience to read through the countless tracts written by political and economic philosophers over the past several hundred years. One after another, these great minds pontificate about natural laws, the social contract, the dialectics of the means of production, and the nature of power, but barely a single word about energy flow and the Entropy Law. It's true that the second law wasn't scientifically formulated until late in the nineteenth century, but that's not really an excuse for letting everyone who came before them off the hook. The ancient Greeks and the medieval Christian scholars hadn't formulated the second law either. But they had intuited it and integrated its central truth into their cultures and their world views.

A lot of political scientists and economists are going to be very unhappy when the emerging entropy paradigm begins to spread its influence into every discipline, including their own. But when that does begin to happen, in the next few years, some very basic concepts that we've long held sacred in our political and economic thinking are going to be radically changed. The process is likely to be thorough for the simple reason that the falsehoods we have for so long entertained are now suspect even among those appointed as the keepers of the faith.

Not every discipline will fare as poorly as political science and economics with the coming of the entropy paradigm. Many anthropologists, for example, have long recognized that the energy basis of a given environment is the primary determinant in the shaping of culture. Anthropologists divide the major periods in history by the changes in how people organized their environment. For this reason, it is important to take a look at the distinguishing features of some of these epochal periods and the common thread that runs through all of them. That thread is the Entropy Law.

PART THREE

Entropy:
A New Historical Frame

History and Entropy Watersheds

Les gens heureux n'ont pas d'histoire. This is an old French proverb which means that 'happy people don't make history.' There is an American proverb which says that 'necessity is the mother of invention.' Put these two sayings together and all of history becomes understandable. Historians will protest. They will argue that the world is much more complicated than that. There are subtleties, nuances, hidden meanings, and unconscious drives to take into consideration. Arnold Toynbee will insist that social history is really a series of cultural and environmental challenges and responses. Oswald Spengler will argue that the history of civilizations is a cyclical process of birth, maturation, and death – like life itself. Ortega y Gasset will chime in with his theory that history is a levelling process in which the tremendous creativity of charismatic minorities becomes taken over and absorbed by the masses and thus made dull and lifeless. Marx will lecture that history is really dialectical and material, and that each unfolding reality contains the seeds of its own destruction and the embryo of a new reality that will replace it.

No need to quibble. Let's just say that all of these gentlemen have identified part of the historical jigsaw puzzle. As with any other puzzle, however, the job of fitting all the separate pieces together is always more difficult when there is no prior knowledge of what the puzzle represents. The key to understanding the puzzle of history is the Entropy Law and these two proverbs. Bring history down to a personal level and things immediately become clear. When you and I are feeling really happy and content with the way our life is going, we rarely if ever entertain the idea of a radical change in the way we go about things. Why should we? As the saying goes, 'Don't knock a good thing.'

On a personal level, we usually begin to think about radical changes in the way we do things when our present approach to life fails us in some way. We have all experienced the feeling of personal crisis, the trauma of having to reexamine our lives, and the fear of trying out something new and unexplored. Yet, it's exactly at these times, when the old way no longer works for us, that we begin to think, sometimes furiously, about ways out. The mental and emotional juices begin to flow, and we start to experiment helter-skelter with various alternatives. Finally, we hit on one or more alternatives that seem to make sense and take hold of them – at least until the next crisis hits.

Personal history is not very different from social history. In both cases happiness marks the blank periods and crises mark the inventive periods. Unfortunately, the exact opposite argument has been advanced by most (not all) modern historians, and for that reason we have to take leave of what our common sense tell us, just for a moment, to examine their thinking on this score.

The leisure, or surplus, theory of history argues that major changes occur in the way people do things when they have built up enough abundance or surplus to allow them the leisure time to think, experiment, and tinker. For example, it's often pointed out that hunter-gatherer societies could never have made the transition to agriculture unless they had first succeeded in building up a surplus. The reasoning is that people always on the brink of famine would 'find it hard to devote resources to a future event – the harvest.'[1] In other words, hungry people aren't going to forgo hunting and gathering for five or six months in order to tend the fields.

This line of argument sounds reasonable at first, but upon closer examination it just doesn't hold up. First of all, let's assume that a hunter-gatherer society did succeed in establishing a surplus. This would mean that their environment was well stocked with more than enough animals, nuts, fruits, and berries to supply all of their needs. Why in heaven's name, then, would they consider uprooting their entire way of life to take up an uncertain, risky, and unexplored new existence tilling the ground? People simply don't destroy their way of life when things are going well – that is, unless they're absolutely crazy. Since we are reluctant to suggest a 'crazy people' theory of history, the fuzzy thinking behind the leisure or surplus theory of history must reside somewhere else. It does. And by now it should come as no surprise that this kind of thinking can be traced right back to the existing machine paradigm. According to the modern world view, history is a steady line of progress in which the surplus of each period provides the margin of free time necessary to invent new tools and technologies which, in turn, result in even more material surplus which frees even more time for the discovery of even more advanced tools and technologies which result in even more surplus and leisure time and so on. The world machine is constantly being streamlined, improved, and enlarged, and our own lives continue to become more secure and comfortable in the process. This is our world view – the way we look at things. No wonder the world around us is becoming so blurred. History has, in fact, unfolded in just the opposite way from how we have been conditioned to think.

All of the evidence suggests that the hunter-gatherers took up farming out of necessity. The game and edible plant life became increasingly scarce, new territories became exhausted, and further geographic expansion became impossible. The crisis of survival dictated experimentation. New ideas were tried out. Gradually, step by step, farming took over as the old hunting-gathering way of life proved less and less economical. Studies of the few remaining hunter-gatherer societies bear out the 'deprivation, crisis, experimentation' thesis. Still, it's not necessary to harp on this one epochal shift in history because we do have records of other major changes in human culture since then, and without exception they show that the great changes occurred not as a result of the building up of abundance but as a result of the dissipation of the existing stock of resources. What this means is that history is a reflection of the second law. The overall entropy process is always moving toward a maximum. With every single occurrence, some amount of energy becomes forever dissipated. In the course of history, critical watersheds are reached when all of the accumulated increases in entropy result in a qualitative change in the energy source of the environment itself. It is at these critical transition points that the old way of doing things becomes inoperative. The entropy of the environment becomes so high that a shift to a new energy environment occurs, along with the creation of a new mode of technology and the shaping of new social, economic, and political institutions.

The Entropy Law also tells us that each of these qualitative shifts in the environment is more harsh and exacting in terms of available energy than the preceding one. This is because, with each successive stage, the stock of available energy in the world has dissipated to a lower and lower level. The overall disorder of the world is always increasing; the amount of available energy is always decreasing. Since human survival depends upon available energy, this must mean that human life is always becoming harder and harder to sustain and that more work, not less, is necessary in order to eke out an existence from a more and more stingy environment. Because there is not enough time in a day for human beings alone to perform the additional work required by the harsher energy environments, more complex technologies must be devised at each stage of history just to maintain a moderate level of human existence.

Upholders of the Newtonian paradigm cannot stomach such thoughts. They argue that new and more sophisticated technologies continue to create greater abundances by replacing less-efficient human energy with more-efficient non-human energy – all of which lessens people's burdens in life. That's what progress is all about. In fact, it's not uncommon to measure cultural progress in terms of the increased use of nonhuman energy. In hunter-gatherer societies, people have to depend largely on their own muscle power as their primary source of energy. An average adult is capable of generating about one-tenth of one horse-power. Compare that figure with the thousands of horsepower or machine power that the average American has at his disposal today as a result of modern technology and it becomes obvious, say the upholders, that history is progress and that people are better off now than in the remote past. Behind this kind of thinking rests an essential assumption: that the greater the energy flow-through, the more efficient a society is, the more progress civilization is making, and the more ordered the world is becoming.

It is now time to dispel such foolish notions once and for all. It is true that each new major development in technique generally speeds up the process of extraction and flow of energy through the system. Remember, though, that energy can never be created or destroyed, and it can only be transformed one way – from available to unavailable. Therefore, every so-called advance in efficiency, as measured by new technologies designed to speed up energy flow, has only hastened the overall process of dissipation of energy and disorder in the world. As the process of energy flow has been speeded up, the period between each new entropy watershed has shortened. It took millions of years to exhaust the environment that supported hunter-gatherer societies before they had to make the transition to an agricultural base. It took thousands of years before people finally 'had to' move from an agricultural to an industrial environment. Within just a matter of a few hundred years people have exhausted the resource base (nonrenewable energy sources) of the industrial environment and today face a new entropy watershed.

Moreover, contrary to the prevailing wisdom, applying more and more energy per individual in order for each person to survive is not more efficient – that is, if efficiency is properly defined as a reduction in work. It is instead quite the opposite. Work, in the final analysis, is nothing more than the using up of available energy. Today, in the modern industrial world, we have to 'use up' a thousand times more energy per person to maintain ourselves than was true a million years ago. If we've deluded ourselves into believing that, just because the work is being done by machines rather than by muscle power, somehow 'less' work is being done, then we are sadly mistaken.

Throughout history, there have been exceptional cultures which have been able to survive for long periods of time within the same energy environment. They were able to make the transition from a colonizing to a climactic stage of existence. For all of these cultures, adapting to the existing environment meant slowing down the flow-through of energy and thus slowing down the entropy increase in the overall environment. Of course, in the real world, it's impossible even under the most adaptive climactic system to forestall an eventual qualitative change from taking place in a particular energy environment. The question is always one of how quickly or how slowly these successive entropy watersheds will be reached. It is interesting to observe that those cultures that have moved into a 'steady state' with their surroundings have tended to see the world as a very closed system, a system they had already filled and from which there was no escape. For them, 'living within their limits' was a matter of second nature.

The modern world view, however, reflects a very different conception. The machine paradigm emphasizes matter in motion. It puts a premium on locomotion and distance. It is bound up with the image of constant growth. Limits are a sign of defeat. The spirit of our age is one of expansion and conquest. Above all, there are always new worlds to conquer. Except now the human population is doubling every forty years, and every nook and cranny on the entire globe appears to be filling up, with standing room only. We find it harder and harder to locate sources of available energy and more and more difficult to find places to discard our energy wastes. We are finally reaching the outer limits of the planet earth, and as we begin to jostle each other back and forth, a new voice can be heard from deep inside the crowd. The voice is getting louder and it's saying that we must learn to 'live within our limits.' The colonizing stage for Homo sapiens is truly over.

Yet, there are those who refuse to accept the obvious. Today, the frontier mentality remains alive and well among space enthusiasts who claim that we can always move on to colonize and exploit other planets. Their expectations can't be met. Sending up just the population increase on earth of six days of births would cost the equivalent of our entire gross national product for one year. Then, too, astronomers tell us that the nearest solar system to ours with planets that might possibly be comparable in climatic conditions is ten light-years away, and with our present technology it would take over a hundred years to travel there. And then there is no assurance that life as we know it could possibly be maintained. Finally, the idea that valuable resources could be mined and sent back to earth from other planets in the quantities needed is completely ridiculous. The cost of mining additional resources on earth is already becoming prohibitive. Even assuming we could locate planets with resources that would be usable in some way here on earth, there is no way we could ever afford the costs of mining and transporting the materials from these distant places.

Only by consciously choosing to respect the physical confines of this closed system we call the planet earth can we make the radical adjustment that is essential for our continuation as a species. Our survival and the survival of all other forms of life now depend on our willingness to make peace with nature and begin to live cooperatively with the rest of our ecosystem. If we do so, and allow the natural recycling process the time it needs to heal the wounds we have inflicted on the earth, then we and all other forms of life can expect a long and healthy sojourn on this planet.

If we steadfastly refuse to make the change and continue on in our colonizing ways, destroying everything in our path, we may find ourselves without a choice in the future. We will eventually reach that critical point where the matter-energy of the planet will be so depleted that even with a complete turn-around to a climactic mode, there will be too little low-entropy terrestrial endowment left to allow the natural recycling process to restore a measure of ecological balance for the continuation of life.

The transition from a colonizing to a climactic mode of existence is the most profound change our species will ever have to make. That crossroads is now before us.

The Last Great Energy Watershed

History does indeed follow the Entropy Law, as a brief survey of two major periods will demonstrate. Because of its relevance to the development of the modern world paradigm and the way we have come to organize our way of life here in the United States, let's take a brief look at western Europe between the fourth century and the nineteenth century A.D. Historians separate this stretch of fifteen centuries into two parts: the Dark Ages and medieval era, and the industrial era. (The Renaissance is generally recognized as a transition time between the two epochs.)

Popular textbooks generally attribute the transition from the medieval era to the modern era to a great reawakening of the human mind – as if for some strange reason all of humanity had decided to stop thinking and simply hibernate for several centuries. While scholars argue back and forth about the significance of the Protestant Reformation, the rise of the bourgeois class and capitalism, and the opening up of trade routes in the great metamorphosis that occurred, few spend much time at all discussing the major underlying cause for the changeover. Between the thirteenth and sixteenth centuries, western Europe experienced an entropy watershed. Wood, the energy base of the medieval way of life, became increasingly scarce. Population pressure further exacerbated the shortages, and the subsequent search for alternatives finally led to the replacement of wood with coal. The move from an energy environment based on wood to one based on coal radically changed the entire way of organizing life in western Europe. The transition from wood to coal was the principal factor behind the demise of the medieval era and the emergence of the Industrial Revolution.

A visitor to Europe today is likely to be impressed with how well used every available space is. Everything appears to be partitioned into neatly arranged geometric sizes and shapes.

Even the open spaces have a measured look to them suggesting that the whole continent was meticulously planned out and sculptured down to the very last detail way back at the beginning of time. It would be difficult to imagine that in the fourth century the continent was a blanket of dense forest stretching from the Alps to the Carpathian Mountains. A bird could fly over the treetops for hundreds and hundreds of miles seeing only an occasional clearing. In some of these tiny open spaces a little smoke might be coming from an open fire. Nearby, there might be a few thatched huts and twenty or thirty people scurrying back and forth near the edge of the forest.

The land in western Europe was very different from the land in the semi-arid Middle East. There the soil was extremely light in composition; in the wetter climate of western Europe the soil was often sticky and heavy, making it much harder to plough. This difference in the environment necessitated some basic changes in cultivation that profoundly affected the future development of the continent.

The old Roman scratch plough was not strong enough to turn over the richer and heavier soils of Europe. By the middle of the sixth century, Slavic peasants began to use a new heavier type of plough with wheels and two blades. The cross plough was equipped with 'a vertical blade to cut the line of the furrow, a horizontal ploughshare, and a mouldboard to turn over the sod.'[2] This new type of plough attacked the soil so violently that traditional cross-ploughing of fields was no longer necessary.

The new cross plough changed the entire organization of agricultural life. Because it was so heavy, the cross plough required a team of eight oxen to move it. Since no single peasant family owned that many oxen, teams had to be used cooperatively. Equally important, with the new cross plough it was no longer practical to fence off land into private strips. The big heavy plough performed best in long open fields. For both these reasons, communal farming became the pattern on most feudal estates in northern Europe.

By the ninth century, the cross plough had been introduced throughout much of the continent. Its unique effectiveness in ploughing rich river-bottom soils led to the deforestation of mile after mile of low-lying timberland, as the expanding population brought more and more acreage into cultivation.

Two other technological developments followed the introduction of the cross plough. In parts of northern Europe, where the soil was richer, increased population pressure began exerting demands for increased crop yields. In response, the traditional two-field system of farming gave way to a three-field rotation. Under the two-field system, half the land was always left fallow in order to renew fertility; with the three-field system, only one-third of the land remained fallow each year. There were several advantages to this approach. First, production was increased by one-third. Second, there was a one-ninth decrease in the amount of ploughing that had to be done. Of course, the short-range advantage in increased yield only resulted in exhausting the soil faster than it would have under a two-field system. By increasing the use of the land, the three-field rotation system speeded up the dissipation of the soil's energy and hastened the entropy process.

The three-field rotation system also made it possible to replace oxen with horses. Horses worked twice as fast but required grain as well as hay to survive. The increased yield of the three-field system provided the surplus oats necessary to maintain a stock of workhorses. Before horses could be effectively used, however, three technological improvements had to be made. By the eleventh century the modern horse collar had been adopted, horse-shoes had been invented, and a method of tandem harnessing had been perfected. These three technological advances allowed horse teams to pull the heavier cross ploughs, thus greatly speeding up the tilling process.

The cross plough, the three-field system, and horse teams greatly increased production on existing land and spurred the opening up of larger tracts of new land for cultivation. The agricultural surpluses of the ninth through the twelfth centuries resulted in a steady increase in population, which in turn resulted in increased pressure to overexploit existing farmland and deforest still more marginal lands for additional acreage. A vicious cycle had set in, the kind that precedes every major entropy watershed. Further refinements in technology were increasing the flow-through of energy, the population, and the entropy process. By the mid-fourteenth century the watershed had been reached. The population had outstripped its energy base. Soil exhaustion and a growing timber shortage were threatening the populations of western and northern Europe. The introduction, in some parts of Europe, of windmills in the twelfth century (and a greater use of water mills) helped bring previously unworkable land into cultivation – but at the expense of further depleting the forests and increasing the population even more.

According to historian William McNeill:

Many parts of Northwestern Europe had achieved a kind of saturation with humankind by the 14th century. The great frontier boom that began about 900 led to a replication of manors and fields across the face of the land until, at least in the most densely inhabited regions, scant forests remained. Since woodlands were vital for fuel and as a source of building materials, mounting shortages created severe problems for human occupancy.[3]

The economic problem was greatly magnified by the expanding population in cities that needed to be fed. Cities had begun to spring up in the eleventh century as trading centres to handle the agricultural surpluses. Now, with population increasing faster than agricultural output, there were no more surpluses to be traded and the cities began to collapse. The whole fabric of medieval economic, social, and political life began to disintegrate. It was at this juncture that a new energy base took hold, one still partially with us today.

In order to comprehend the magnitude of the medieval energy crisis, it is important to understand how crucial wood was to the life of the times. Like fossil fuels today, wood was used for just about everything. Lewis Mumford drew up a list of some of the particulars:

The carpenters' tools were of wood but for the last cutting edge: the rake, the oxyoke, the cart, the wagon, were of wood: so was the wash tub in the bathhouse: so was the bucket and so was the broom: so in certain parts of Europe was the poor man's shoes. Wood served the farmer and the textile worker: the loom and the spinning wheel, the oil press and the wine presses, and even a hundred years after the printing press was invented, it was still made of wood. The very pipes that carried water in the cities were often treetrunks: so were the cylinders of pumps ... the ships of course were made of wood and ... the principal machines of industry were likewise made of wood.[4]

Mumford sums up the importance of wood to the life of that period by observing that 'as raw material, as tool, as machine, as utensil and utility, as fuel, and as final product wood was the dominant industrial resource.'[5]

While the clearing of forests for cultivation greatly reduced the available wood supply, it was the quickened pace of commercial activity that led to a timber famine. For example, the new glass works and soap industry required large amounts of wood ash. But it was the production of iron and the building of ships that made the greatest demands. By the sixteenth and early seventeenth centuries, the timber crisis was so acute in England that royal commissions were set up to regulate the cutting down of forests. The regulations proved ineffective. In the 1630s wood had become two and a half times more expensive than it had been in the late fifteenth century.

The answer to the wood crisis was coal. But it was not just a simple matter of replacing one energy base with another. The cultures of Europe had been thoroughly integrated into a wood-based existence. The changeover necessitated the radical uprooting of an entire way of life. The way people made a living, the way people got around, the way people dressed, the way people behaved, the way governments governed – all of this was turned inside out, then upside down.

It started in England in the thirteenth century under Henry II. The people of Newcastle were without firewood and literally freezing to death. The king consented to the mining of coal as an alternative energy source.

In the fifteenth century, Pope Pius II wrote that while in Scotland on a visit he was surprised at the sight of people in rags lining up at church doors to 'receive for alms pieces of black stone with which they went away contented. This species of stone they burn in place of wood of which their country is destitute.'[6] By 1700 coal had begun to replace wood as the energy base for England. Within 150 years the same held true for much of western Europe.

Today we think of the substitution of coal for wood as a great leap forward, a singular triumph for the forces of progress. It would have been difficult to convince the folks back then. Coal was treated with contempt as an inferior energy source. It was dirty and created a great deal of pollution. In 1631 Edmund Howes lamented that 'the inhabitants in general are constrained to make their fires of sea-coal or pit coal, even in the chambers of honourable personages.'[7]

Coal was also more difficult to extract and process than wood. It required the expenditure of a great deal more energy to transform it into a usable state. The reason is to be found once again in the operation of the second law. The available energy in the world is constantly being dissipated. The more available sources of energy are always the first to be used. Each succeeding environment relies on a less available form of energy than the one preceding it. It is more difficult to mine coal and process it than it is to cut down trees. It's still more difficult to drill and process oil, and even harder to split atoms for nuclear energy. Richard Wilkinson, in his book *Poverty and Progress*, reviews the history of human economic development:

During the course of economic development man has been forced over and over again to change the resources he depended on and the methods he used to exploit them. Slowly he has had to involve himself in more and more complicated processing and production techniques as he has changed from the more easily exploitable resources to the less easily exploitable . . . In its broadest ecological context, economic development is the development of more intensive ways of exploiting the natural environment.[8]

Wilkinson's thesis is quite correct, although difficult for most of us to accept. We are used to thinking of great leaps forward in history occurring because someone came up with a better way of doing things. Actually, these so-called better ways are in reality only different ways of doing things occasioned by the need to readjust to harsher, less easily exploitable energy environments. And, as Wilkinson suggests, each new way of doing things ultimately requires the expenditure of more work (or energy) than the previous ways – although the work is carried out by nonhuman energy sources. The development of the steam engine is a good case in point.

When we study the Industrial Revolution in school we're taught that one day a bright young man named James Watt was tinkering in his garage or wherever he tinkered and he came up with a little invention which he called the steam engine. The announcement was made to the world and within the blink of an eyelash the Industrial Age was off and clanking. Nowhere are we told that the ancient Greek, Hero, had devised a steam engine in the third century B.C. but that it was only used as a toy for the amusement of the royal court. There was no thought of using it for the purpose of doing work because they had plenty of slaves for that. The Greeks aside, the story behind the development of the modern steam engine requires knowing because it vividly demonstrates that major technological changes (not just refinements in existing technologies) follow changes in the energy environment.

The modern steam engine was designed and first used to facilitate the mining of coal. As mines had to be sunk deeper into the ground to extract available supplies, it became more difficult to ventilate the mines and to lift the hewn coal up the shafts. During the seventeenth century, mines faced still another problem. At a certain depth the water table was reached and drainage became a priority. All of these problems required a technological solution. The steam engine was the answer. The first steam pump was patented in 1698 by Thomas Savery.

The steam pump used in mining was only the first in a long series of mechanical and structural innovations to come directly out of the new coal environment. For example, no sooner had the problem of mining the coal been resolved with the introduction of the steam pump than a second and equally important problem arose – how to transport the coal to markets throughout the country. Because of its heavy bulk, coal could not easily be transported over land by horse-drawn wagons. English roads were for the most part unsurfaced. The sheer weight of the coal wagons created giant ruts in the roads, making it virtually impossible to travel during rainy periods when roads turned into muddy ditches. At the same time, the cost of maintaining transport horses became increasingly expensive. With farmland in critically short supply, it was not possible to grow food for both horses and people. The answer to the transportation crisis was the invention of the steam locomotive and railroad tracks. Like the steam pump, the steam locomotive was a direct technological response to the needs created by the new coal environment. Together, the steam pump and the steam locomotive laid the technological base for the industrial era that followed.

The steam pump and steam locomotive were much more complex energy-consuming technologies than the axe, horse, and cart of the wood era. But then, the energy environment was more exacting. Throughout history, qualitative changes in technology have always been toward more complexity and greater energy expenditure, because each major change in the environment has been toward less available, harder-to-reach sources of energy.

Not only has more work been required with each new environment, but the new way of doing things is usually perceived as an inferior substitute for the old way. Sometimes this perception is immediate, and sometimes not until well after the substitute has taken hold. Take, for example, canned and packaged food. Very few people today, if left with the choice of having processed food or fresh natural food, would choose the former, although for a long time it was touted as a superior substitute. In the case of processed foods, the amount of energy (or work) needed to produce the product is much greater than the amount used up by the older way.

Wilkinson offers still another striking example: clothes. In prehistoric times, people used leather to clothe themselves. As animal hides became more and more scarce they were forced to replace them with wool from sheep. By the seventeenth and eighteenth centuries, the population pressure on available farmland in Europe made sheep grazing less economical. 'Sheep devour people' became a favourite slogan of the time, and demands were made to turn more grazing land to crop cultivation. This required a substitute for wool. The answer was found in cotton, which could be grown cheaply in the overseas colonies and imported back to the mother countries for conversion into cloth. People were not exactly overjoyed with the new substitute, as Friedrich Engels pointed out in his book *The Condition of the Working Class in England*:

The working classes very seldom wear woollen clothing of any kind. Their heavy cotton clothes, though thicker, stiffer and heavier than woollen cloth, do not keep out the cold and wet anything like the same extent as woollens ... Gentlemen, on the other hand, wear suits made from woollen cloth, and the term broadcloth is used to designate the middle classes.[9]

Similarly, today, we are forced to rely more and more on synthetic fibres, but if given the choice many people would prefer clothes made of 100 per cent cotton, or wool.

Though less desired, each successive clothes substitute has still required the expenditure of greater work (energy) to produce it than its predecessor. It didn't take a great deal of work to kill an animal, tan its hide, and fashion clothes for an entire family. The feeding and grazing of sheep, the shearing and weaving of yarn, and the sewing of woollen garments meant that a great deal more human and nonhuman energy had to be put into the process. With the growing and processing of cotton, even more energy is expended. By the time we consider synthetics, the whole chemical process – beginning with the drilling of oil and leading up to the giant factories stamping out the final designs – requires an expenditure of work (energy) per garment that is astronomical compared with the killing of animals and tanning of hides.

This is what we call 'progress.' In the next two chapters we will explore in much greater detail some of the specific ways that changes in technology increase the use of work inputs per unit of output, hastening the entropy process and the buildup of disorder in the world.

Technology

'The emperor isn't wearing any clothes!' This is the way one feels at the first recognition of what technology really is. Remove all of the mystique that surrounds it, and what is left, naked and exposed, is a transformer. Every technology ever conceived by the genius of humankind is nothing more than a transformer of energy from nature's storehouse. In the process of that transformation, the energy flows through the culture and the human system where it is used for a fleeting moment to sustain life (and the artifacts of life) in a nonequilibrium state. At the other end of the flow, the energy eventually ends up as dissipated waste, unavailable for future use.

It is ironic that as technology has become more complex and has enlarged its domain in the world, we have come to see it as something independent of nature, as if it were generating its own energy from scratch or, through some mysterious process, were adding to the existing energy source to get more out of it than was there in the first place. The fact is, technology never creates energy; it only uses up existing available energy. The larger and more complex the technology, the more available energy it uses up. As awesome and impressive as our technology might sometimes appear, it too operates under the supreme reign of the first and second laws, just like everything else in nature. Those laws again: first, all matter-energy in the world is constant; it can be neither created nor destroyed but only transformed from one state to another. Second, the transformation of energy is always from an available to a dissipated form, or from an ordered to a disordered state. Technology is the transformer – nothing more, nothing less.

Even though all of this is rather obvious, we still continue to live under the delusion that our technology is freeing us from dependence upon our environment, when nothing could be further from the truth. Life is not a closed system. Human beings, like all other living things, can only survive by exchanging with the environment. Without a constant flow-through of energy from the environment we would all perish within days. Technology makes us more dependent upon nature, even as it physically moves us further away from it; we have become more dependent as we have required increasing doses of nature's energy to sustain our cultural patterns and our personal life-styles.

We also entertain the belief that technology is creating greater order in the world when, again, the opposite is the case. The Entropy Law tells us that every time available energy is used up, it creates greater disorder somewhere in the surrounding environment. The massive flow-through of energy in modern industrial society is creating massive disorder in the world we live in. The faster we streamline our technology, the faster we speed up the transforming process, the faster available energy is dissipated, the more the disorder mounts.

In short, we live in a kind of nightmarish Orwellian world. We have convinced ourselves that the way we go about things is creating a world quite different from the one we are really making. Just as in Orwell's *1984*, where society was convinced that war was peace and lies were truth, we have come to believe that disorder is order, that waste is value, and that work is nonwork.

The more our world slips deeper into chaos, the less willing we are to identify the source of the problem. Instead, we wrap ourselves up even tighter in our technological garb, defending it against all criticism, unable to acknowledge what it is doing to the environment we live in, and even less able to acknowledge what it is doing to us. We continue to cling to the fiction that we are securely clothed and protected, even as we become more exposed and endangered by the disordered fragments of a world of our own making.

External Costs

It has become fashionable of late to talk about the 'external costs' of technology. The term is used to refer to the unanticipated costs that arise as a result of the so-called secondary effects produced by a particular product, process, programme, or service. Everyone is becoming more and more familiar with external costs. When a nuclear power plant breaks down and spews low-level radiation into the environment, the question immediately arises as to who should pay for whatever damage is done – the public, the utility, the designer, or the government. This is referred to as an external cost. Whenever politicians or economists talk about external costs, they convey the feeling that what is involved are the nuisance-causing side effects that sometimes accompany technologies. These side effects are often costly, but they are tolerable and absorbable because the benefits derived are always considered greater than the external costs generated. This just isn't so.

'External costs' is just a convenient dodge to try and avoid the consequences of the Entropy Law. The disorder created by each new technology is not a side effect. Nor is it less costly in the long run than the benefits derived from the particular technology. If this were so, then technology could claim to violate the second law and truly deserve the sacred status we have given it. The truth is, every technology creates a temporary island of order at the expense of greater disorder to the surroundings. Twenty years ago, no one in America would have been willing to believe this. Back then we all believed that the benefits of technology always outweighed the harm. If technology sometimes failed or produced unfortunate side effects, then the solution was to be found in the application of new technologies to cover up the mistakes of the old. Today, tell people that a new technology is going to be introduced that will be of great benefit to them and society and their immediate reaction is likely to be one of scepticism. Whether it be a new government programme, a new way to harness energy, or a new superdrug, the response is often, 'Let's wait and see.' While on the surface the benefits might appear worthwhile, there is always the doubt gnawing away at each of us that whispers: 'I don't know when, or where, or how, but sooner or later this, too, is going to screw up and create greater problems than what I'm going to get out of it.' Nuclear power – low-level radiation and cancer. Bigger and faster automobiles – carbon monoxide poisoning and pollution. The Entropy Law is not selective. It works everywhere, all of the time. According to Jacques Ellul, the author of *The Technological Society* and perhaps the most renowned critic of technology, 'History shows that every technical application from its beginning presents certain unforeseeable secondary effects which are more disastrous than the lack of the technique would have been.'[10]

The next time a technician, politician, or businessman tells you he or she can eliminate the secondary problems associated with a particular programme, product, or process, with better planning or better leadership or better design, think about the second law. It is true that the secondary disorders caused by a particular technology can be temporarily solved by the application of additional technology. But the solution will inevitably result in even greater disorders than the one it solved. Again says Ellul, 'Every successive technique has appeared because the ones which preceded it rendered necessary the ones which followed.'[11] That's the second law and there's no way around it. Still we wonder why it is that the more we technologize the world around us, the more things seem to malfunction and fall apart.

The world is becoming more disordered because each time we apply a new and more complex technological solution to a problem, it's like dousing a fire with petrol. The faster we multiply the 'transformers,' the faster the available energy is used up, the faster the dissipation and disorder that result. The problems proliferate faster than the solutions.

There are those who would argue that every culture throughout history has used technology and, to a greater or lesser extent, has been able to adjust to it without catastrophic consequences. So why shouldn't we be able to do the same? What they fail to recognize is the key difference between technique in the modern culture and technique in cultures that have preceded us. In all the other civilizations before the industrial era, technique was limited in the functions it performed. It was a tool but not a way of organizing life. In the world machine paradigm, technique has become the way of organizing all of life's activities. A conscious attempt is made to bring technique into every aspect of our daily existence. The goal is predictability and synchronization. As long as parts of our culture remain outside of the technical process, they remain unpredictable and uncontrollable. It is argued that the system as a whole can never really function smoothly until these pockets of unpredictability are eliminated.

In our efforts to bring technique and order to all of life's activities, we are merely speeding up the transforming process and thus hastening the entropy process. Scientist Eugene Schwartz, in his book *Overskill*, compared our efforts to create the technological society to a giant squirrel cage 'wherein technicians must run faster and faster to remain in the same place. Unlike the squirrel cage, however, the faster they run, the further they fall behind. Each quasi-solution has a multiplier effect on the residue of problems.'[12] Moreover, each new set of problems is more difficult to solve than the ones that came before, because with the passage of each occurrence, the entropy of the environment has increased and the available free energy has decreased. It becomes harder to maintain order and more costly to generate order. The more we try to spread technique over the culture, the more fragmented society becomes. The whole process of increased complexity, increased problems, increased entropy, and increased disorder proceeds exponentially, and that's what makes the modern world crisis so frightening.

Lest we slip over that word *exponential* without much thought, just consider what it really means relative to the problems of a technological society. Exponential growth is a doubling process. According to ecologist G. Tyler Miller, if you were to take a piece of paper (about 1/254 inch thick) and double its thickness just 35 times, it would extend the distance from Los Angeles to New York City. If you doubled it 42 times it would reach from where you're sitting to the moon. If you doubled it just a few more times, say to just over 50 times, the thickness of that paper would extend 93 million miles, from you to the sun. The exponentiality of the technological fix is a one-way ticket to disaster for life and for the planet earth.

Diminishing Returns of Technology

American business magazines are full of articles bemoaning the decline in American technology. This is of great concern because advances in technology have been the mainstay behind American economic superiority in the world. There are various reasons given for the drop-off. Some blame it on inferior academic preparation. Others blame it on falling profit margins and disincentives to invest. It is argued that government regulations and restrictions extend the time between the discovery of new products and processes and their introduction into the marketplace, making research both costly and risky. Still others blame the downturn on environmental standards which restrict the way research can be conducted and the way the final product can be used.

In hearings before the Joint Economic Committee of Congress in 1976, many of the experts on technology even suggested that diminishing returns might have set in across the board and that America's great technological strides of the past would probably never be repeated. One witness before the hearings shocked the congressional assemblage by pointing out that in the past ten years, after all of the billions of dollars spent in research and development, only two technological breakthroughs with a 100 per cent market potential were introduced – permanent-press pants and pocket calculators. Not very impressive. With all of the bickering and bantering over who or what is to blame and what can be done, very few of the experts seem to have understood the underlying reason behind the diminishing returns in technology.

Back once again to basics. Technology is not some kind of independent, autonomous force. It is merely a transformer of energy. Therefore, major breakthroughs in technology come on the heels of qualitative changes in the energy source. This is because specific modes of technology are designed to transform the energy of specific types of energy environments.

(For example, remember that the steam engine, which was the basic technology for the Industrial Revolution, was invented to transform coal energy from beneath the earth's surface.) During the early stages of a new energy environment, the new mode of technology expands in every direction. This is the time period in which the technological base of the new energy environment is being laid. There is plenty of experimentation, and a wide range of technological derivatives are created one after the other. Often the unit cost of the derivatives actually becomes cheaper and cheaper with further refinements in the technology. However, as the mode of technology proliferates and the energy flow increases through the system, the entropy of the environment steadily moves toward a maximum, and diminishing returns begin to set in all along the line of energy flow. Sucking more energy out becomes more expensive and complicated. The disorder created by the past flow-through accumulates, exerting increasing pressure, and putting further restraints on new technological possibilities. A critical point is reached when the existing 'type' of technology can no longer sustain the level of energy transformation that the society has come to depend on during the earlier stages. From this point on, less effort is put into new technological ideas and more devoted to readjustments in existing technologies in an attempt to both solve the problems caused by increasing disorders to the overall environment and at the same time meet energy demands in the face of a depleted energy base. This is exactly what's happening in the United States today.

According to Mobil Oil chairman Rawleigh Warner, Jr., 'Industry has been compelled to spend more and more of its research dollars to comply with environmental, health and safety regulations – and to move away from longer term efforts aimed at major scientific advances.'[13] Some industries, like iron and steel, are now using up over 20 per cent of their total capital expenditures on pollution-control equipment.[14] A study done by the Brookings Institution concluded that such expenditures accounted for a 17 to 20 per cent decline in the US economic growth rate in one recent year. The same study estimates that by the mid-1980s annual business ex-

96

penditures for environmental related adjustments will exceed $40 billion (in 1975 dollars).[15] At the same time, the National Petroleum Institute estimates that as a result of the depleting energy base, it will take an additional $172 billion in the years just ahead to explore for and process less easily exploitable sources of fossil fuel energy.[16]

Eventually the technology bottoms out altogether as the energy environment it was made for nears its own entropic watershed. In a recent cover story in *Newsweek* entitled 'Innovation,' the editors candidly acknowledged this basic reality: 'To some extent, of course, erosion in America's technological edge is inevitable. No longer can the US count on the natural abundance of its frontiers . . . the resources have been explored and sometimes depleted.'[17]

It should be emphasized that all along the line of a particular type of energy flow, the dominant techniques and technologies used conform to a common energy bond. The kind of economic institutions, the form of transportation and communication systems, the location, design, and operation of cities and towns are all derived from a common energy flow. When that energy flow reaches an entropy watershed and a new energy environment takes over, the various technological forms that served the old energy flow are either radically changed or, more often than not, allowed to simply atrophy as the old energy spigot runs dry. Even a casual examination of the technological and institutional changes that took place when society moved from a wood to a coal energy base and from a coal to an oil energy base bears out this simple observation.

Institutional Development

Historians have observed that at a certain stage in the development of a culture or civilization, a process of universalization sets in. That is, there is a concerted attempt to consolidate the various economic and political activities under more centralized control. Eventually a point is reached where it is impossible to further consolidate and the culture or civilization begins to break down and fragment. But before that point is reached, each succeeding crisis is met by an escalation in centralized control. Certainly, that has been the case with all of the industrializing nations. Each new social or economic crisis is inevitably dealt with by the establishment of some new form of control or regulation, and a greater measure of authority is placed in fewer and fewer hands. Rarely does a crisis get resolved by decentralizing power and placing responsibility and control with a greater number of people. The tendency of institutions and processes to become larger, more complex, and more centralized is the same tendency we see with various forms of technology. The reason for this can be found in the operation of the Entropy Law.

Economic and political institutions, like machinery, are transformers of energy. Their job is to facilitate the flow of energy through the culture. During the first stage of a new energy environment, the economic and political institutions are generally more flexible. This is because, in this early stage, the energy flow is being used primarily to create the new transformers (technologies) for the new energy environment. While some energy is flowing through the larger systems, much of it is still lavished on the creation of the transforming machinery. At this stage, the economic and political institutions serve more as designers and coordinators. Their role is an innovative one and therefore requires a great deal of manoeuvrability.

Even in the case of societies that place a heavy emphasis on

98

regimentation of the population during the takeoff stage of technology development, the economic and political institutions still have to remain relatively flexible and uncomplicated in order to take advantage of all of the technological possibilities that need to be explored and experimented with.

In the takeoff stage of a new energy environment, the population at large is always deprived of a great deal of the energy flow because it is being diverted into the building of the new base of energy transformers. The harsh living conditions and the regimentation are generally tolerated because the population is desperate, having experienced the tremendous deprivation and disorder that accompanied the last stages of the old energy environment. Certainly this was true for many millions of European peasants forced off the land and into the urban slums in the seventeenth, eighteenth, and nineteenth centuries. As bad as conditions in the factories and industrial towns were, they were no worse than the conditions in the countryside where timber shortages, exhausted soil, and overpopulation were producing starvation and panic.

In the second stage of a new energy environment, more and more energy begins flowing directly through the social system. At this stage, the initial technological base is in place, and from here on in the technology begins to multiply. It is at this point that the weight of the entropy process begins to exert itself. The flow-through of energy creates greater secondary disorders all along the flow line.

The disorders are of three general kinds: those precipitated as a result of the transformation of energy into various products or services; those resulting from the exchange of energy between individuals and groups; and those resulting from the discarding of energy wastes. As mentioned earlier, people depend upon energy flow for their survival and are continually involved in the process of transforming, exchanging, and discarding energy. We work for a living, we buy things, and we throw things out or exchange them for other things. This is what the energy flow line and economic life are all about. Every time we add our own labour to a product or perform a service we expend energy and increase the overall entropy of the environment. Every time we exchange money for a product or a service, the legal tender we use represents payment for previous energy that we expended. Money, after all, is nothing more than stored energy credits. Salaries and wages represent payment for work done or energy expended. Every time we discard something, whether it be an old coat or yesterday's leftovers, energy is again being dissipated. At every stage in the flow line, energy is being transformed, exchanged, or discarded. In the process, energy is always being dissipated and the entropy of the environment is always increasing. The type, scope, and magnitude of the disorders, then, depend on how the flow line is set up. The way the work in a society is allocated (the transforming of energy), the way the energy is divided up between various people, groups, and constituencies (the exchange of energy), and the way the wastes are disposed of at each stage of the flow-through process (the discarding of energy) determine the social, economic, and political nature of the disorders that surface.

As disorders all along the flow line mount, the flow itself becomes impeded. To maintain maximum flow-through it becomes necessary to continue to reorder the disorder that is proliferating faster and faster in every part of the system. Economic and political institutions enlarge their functions and expand their outreach. They begin to serve a maintenance and repair function. The bureaucracies grow with each succeeding crisis. When the disorder at any point along the flow line becomes so great that it threatens the continued functioning of part or all of the society, it moves the appropriate institution to remove the blockage. The institutions swell as they are forced to absorb and contain the mounting social and economic disorders and maintain the maximum flow-through. Periodically, the institutions burst as they can no longer manage to hold back the escalating disorder. When this happens, new, even larger and more centralized institutions are designed to reorder the disorder and so on.

At the same time, the state attempts, if at all possible, to extend its domain into new geographic territories in order to obtain new sources of available energy to compensate for the depletion of its existing stocks. All imperial or colonial expansion is designed to secure new sources of energy. Of course, new conquests require the expenditure of even more energy to provide for armies, weapons, and bureaucracies to occupy and administer the new territorial possessions. The state's institutions become even more enlarged and centralized.

Finally the society reaches stage three, where its institutional complex (its economic and political organization) is so centralized and enlarged that it takes more energy to maintain it than the system can afford. Anyone who has looked at the escalating costs associated with the maintenance of giant multinational corporations and huge government bureaucracies can't help but notice that more and more energy (or work) is expended operating them, while less and less work is got out of them. The institutional complex, which was supposed to facilitate the flow of energy through the culture, becomes a parasite, sucking up much of the remaining energy source. All along the energy line the flow slows up, and the society begins to atrophy. In the final stages, even the institutional complex can't be maintained with the energy it needs from the environment. At this point the entire complex begins to disintegrate. The society becomes increasingly vulnerable to conquest by other nations or internal upheaval and revolution. An entropy watershed has been reached. The rise and fall of ancient Egypt, Mesopotamia, Rome, and the hydraulic civilizations of the Far East are all classic illustrations of how this whole process works. But every other major civilization that we know of has also followed the same course.

Does it always have to be this way? Will humankind be forever trapped in this pattern of social development? The answer is yes, as long as most human societies choose to remain in a colonizing stage rather than move into a climactic stage of existence.

In a colonizing stage, the emphasis is always on increasing flow-through. As we have already seen, increasing flow-through always hastens the entropy process, speeding up the accumulation of disorder, which results in greater institutional control, complexity, centralization, and all of the other

101

things just mentioned. It's no accident, by the way, that the term *colonization* is used. The colonizing of overseas territories by the great imperial powers in the eighteenth and nineteenth centuries is a graphic example of the truth of the historical development thesis just outlined. Colonial administrations were designed to increase the flow-through of energy from the colonies to the mother country. As the entropy process proceeded and disorders mounted, the colonial administrations had to enlarge their bureaucracies and their armies, diverting more energy away from the flow line and toward their own maintenance. Finally, the armies of occupation and the colonial administration could no longer be maintained by the mother countries and became so parasitic on the local energy supply that the native populations rebelled and overthrew them.

Only in a climactic framework can the process of increasing complexity and centralization be slowed down. By minimizing the energy flow-through, the entropy process slows down (it can never be stopped) and the disorders slow down as well. If the energy flow-through is maintained at a low constant level, the institutions responsible for transforming it through the social system can be maintained in a steady low-growth state. Only when a society attempts to get more and more energy out of its environment do the institutions (and other technologies) grow concomitantly in both complexity and control. A climactic existence, then, favours small, decentralized institutions; a colonizing existence favours large, centralized institutions.

Specialization

Specialization goes hand in hand with increased complexity and centralization. In a technological society everything becomes a component of the expanding social machine, including human beings. As the overall functions of a society become more complex and centralized, each individual function becomes more refined, more limited, and more dependent on every other function in the system for its survival.

Anyone who knows anything at all about machinery will be quick to tell you that the simpler the machine and the fewer the parts, the fewer things can go wrong. Simpler machinery is also more flexible and can be adjusted more easily to changing needs. In contrast, our technological society has become so specialized by function that whenever any part of the machinery malfunctions the entire system threatens to break down.

On November 9, 1965, some 30 million Americans experienced firsthand what can happen in an overly specialized society when just one tiny specialized function goes on the blink. On that day, late in the afternoon, a small electrical relay in a power station in Ontario, Canada, failed. Within minutes, almost the entire northeastern portion of the United States was without electricity. Thousands of people found themselves stranded in elevators and subway cars. The traffic lights all went off, creating massive traffic jams up and down the East Coast. As evening set in, that part of the nation was plunged into total darkness – without lights, without heat, and without many of the other accoutrements of the technological society that we have come to depend on for our very survival.

We find less dramatic but just as poignant demonstrations of the vulnerability of our overspecialized society in our everyday experience. For example, if a steel strike in Gary lasts too long, a person may be laid off from work even though he is a cashier at a department store in Denver. Without steel, the auto plants can't produce cars. If the Big Three auto makers shut down or curtail production, then ancillary industries that provide autos with everything from plastic to glass must curtail their own operations. With one out of every six jobs directly or indirectly related to the auto industry, within weeks of a slowdown or a shutdown the economy would begin to slump. There would be a fall in consumer purchasing power, and the cashier at the department store in Denver might find himself in the unemployment line as a result of fewer purchases being made.

Technological specialization so limits the scope of operation of each function in society that it is virtually impossible to readapt a particular function to perform a different task. Every component is designed specifically to perform the task given it and no other. If the nature of the task changes, the component becomes worthless. This is equally true in the case of human specialization. Every day we encounter the limitations of expertise. For example, a foot doctor knows only about feet and cannot be expected to give sound medical advice about any other physical ailment. An antitrust lawyer only knows antitrust law and can't be expected to know about divorce law. A geneticist knows only about genes and chromosomes and can't be expected to understand the workings of a jungle ecosystem. The *Dictionary of Occupational Titles* now lists over 20,000 specialized jobs in America. We have come to the point where each of us knows more and more about less and less, until as a society we all know almost everything about nothing.

Overspecialization, say the biologists, is one of the most important contributing factors in a species' becoming extinct. When a species becomes overspecialized in a particular type of ecosystem, it is usually unable to adapt to a change in environment. It does not contain the flexibility and diversification to enable it to make the transition. The same holds true with human society. Today we have become so overly specialized and adapted to the existing energy environment that we have lost much of the flexibility required to make a transition into a radically new energy mode.

World Views and Energy Environments

This entire discussion now brings us to these central questions: Why did the Newtonian world machine take hold when it did? Why is it still with us today? Why is it only now, a hundred years after the formulation of the second law of thermodynamics, that entropy is emerging as a competing paradigm?

As we have already seen, the basis of all of life is energy, ultimately derived from the sun. Technologies and institutions have served, throughout history, as transformers, facilitating the flow of energy from the environment through the human and social systems. The kinds of specific technologies and institutions that human cultures have developed have been a reflection of the kinds of energy environments they have lived in. This is so because different energy environments require different types of transformers. What should now be equally obvious is that the kind of world view a culture or civilization develops is also a reflection of the particular energy environment it finds itself in. A world view provides an explanation for why people organize life's activities in a certain way. When energy environments change, people are forced to change their ways of doing things – namely, the ways in which they transform energy from the environment. When people make these basic changes in the ways they relate to the world, their world view changes to reflect, rationalize, encourage, and explain the new circumstances.

That doesn't mean that only one kind of world view can emerge from one type of energy environment. In fact, similar energy environments have spawned different world views. However, whatever world view emerges must be at least compatible and consistent with the energy environment it interacts with. The various world views of different hunter-gatherer cultures would all be totally useless in any agricultural setting, just as all agricultural world views would be totally out of place in any advanced industrial environment. The energy environment, then, establishes the broad limits within which human beings then make choices over the kind of belief systems they adopt.

Take, for example, the Newtonian world machine paradigm. It emerged in the seventeenth and eighteenth centuries as European cultures were in the process of shifting from an environment based on renewable energy resources to one based, for the first time in history, on nonrenewable forms of energy. With the shift to nonrenewable energy forms humankind moved from a world of cycles and flows to a world of quantities and stocks. The world views shifted just as radically.

Cultures that were organized around the transformation of renewable energy sources perceived the world as a continual coming and going of seasons. The cycles of birth, life, death, and rebirth were qualitative processes. The energy sources were full of life and colour. With renewable resources, the concept of order and decay was an ever-present reminder of the way the world unfolds. The world views of the ancient Greeks and the early Christians reflected the realities of an energy environment based on living, renewable energy sources.

Contrast the features of renewable energy sources with nonrenewable. Coal and oil are lifeless quantities. They can be divided and redivided and still the individual parts will contain the same attributes as the whole. A speck of coal is little different in composition from a chunk of coal, while the leaf of a plant is very different from the stem or roots. Nonrenewable resources represent a fixed stock. They can be easily quantified. They are subject to precise measurement. They can be ordered. Renewable resources, on the other hand, are forever changing and flowing. They are never still. They are always in the process of becoming. They are hard to subject to precise measurement. With its mathematical formulas, its emphasis on measurement, and its concern with location and distance, the Newtonian world machine paradigm was tailormade for effectively harnessing an energy base of nonrenewable resources.

Scholars have often wondered exactly why the notion of unlimited progress took hold alongside the idea of the world as a machine. The answer is also to be found in the nonrenewable energy base. Here for the first time was a gigantic, seemingly endless stock of stored solar energy – 3 billion years' worth, to be precise. As society hungrily dipped into this storehouse of energy, the concept of cycles and seasons fell further and further into the background. With this bonanza of billions of years of stored solar energy, there was no longer any need to wait for the sun to come up each day and shine upon us, creating energy and life. We had all the energy we needed to replace the sun and would never again have to wait for nature to take its course. Time, then, lost its connection with the natural unfolding of things. Time became a function of how fast we could harness the stored solar energy that lay deep in the coal seams and oil reservoirs. Is it any wonder, then, that under the Newtonian paradigm time can be speeded up and slowed down and turned back and then forward?

With nonrenewable energy we could turn the sun off and on at will. We could make the sun stay out twice as long if we chose because we were dealing with the 'stored sun' – sun that we could take out of the ground and manipulate at will. With nonrenewable energy sources people became increasingly convinced that they were no longer dependent upon nature, and that they could reorder the world to their own making. We no longer had to concern ourselves with the idea of dissipation, decay, and disorder. Time could be controlled, energy could be created, and material progress could be assured.

The Newtonian world machine provided the rationale for this new way of looking at life and organizing its activities. It is being challenged now and will soon be abandoned, because we are about to make the transition away from a nonrenewable energy base toward renewable sources of energy once again.

PART FOUR

Nonrenewable Energy and the Approaching Entropy Watershed

The Energy Crisis

Addiction! There is simply no other way to accurately describe America's energy habit. The statistics are overwhelming. With only 6 per cent of the world's population, the United States currently consumes over one-third of the world's energy.[1] Even the energy consumption in other highly industrialized nations pales in contrast to our own. In Sweden and West Germany, for example, per capita consumption of energy is only half that of the United States even though their standard of living is comparable to ours.[2] The United States consumes more energy per year than all the countries of Western Europe combined, even though their population exceeds ours by 75 per cent.[3]

Although it's impossible to grasp the sheer magnitude of our energy flow-through, consider just the statistics on electricity generation. In 1970, the United States generated 1·7 trillion kilowatt hours of electricity from oil, natural gas, coal, and nuclear sources. This was more than was produced in the world's four other greatest consuming nations combined – the Soviet Union, Japan, West Germany, and Great Britain.[4]

When US energy consumption is contrasted to poor Third World nations, the figures become so great that comparisons become practically impossible to make. How does one even begin to calculate relative energy advantage or deprivation knowing that in a country like Haiti the energy consumption per capita is equivalent to 68 pounds of coal per year, while the per capita consumption in the United States is equivalent to 23,000 pounds per year?[5]

Despite the enormous disparities in energy consumption that already exist, America's need for increased energy is expected to double in the next twenty years.[6] If this exponential growth rate continues, it is estimated that within just 200 years every square inch of available land in the entire United

States would have to be taken up by power plants just to meet the increased energy demands.[7]

Worldwide energy needs are expected to quadruple by the year 2000.[8] This is largely a result of runaway population growth. The population statistics are staggering. Every day 333,000 new babies are born on the planet. Even allowing for 134,000 deaths per day, the net increase in world population is now 200,000 every twenty-four hours. That's 73 million more people in the world next year – all of whom require inputs of available energy to survive.[9]

The population problem can only be grasped by placing it within a historical perspective. It took 2 million years for the human population to reach 1 billion. The second billion took only a hundred years. The third billion took only thirty years, between 1930 and 1960. The fourth billion took only fifteen years. Between 1960 and 1975 the world's population grew at a rate of 2 per cent per year, going from 2½ billion to 4 billion people. At current annual growth rates of 1·7 per cent, the world's population will double once again to 8 billion by the year 2015 and to 16 billion by the year 2055.[10]

Exponential population growth is exerting a tremendous strain on the world's energy base. According to a United Nations study conducted by the Nobel Prize-winning economist Wassily Leontief, in order to meet a moderate rate of global growth over the years ahead, it will be necessary to increase the consumption of common minerals by fivefold and food consumption by fourfold.[11]

What's even more remarkable is the projection made by many international economists that in order to accommodate the minimum needs of the expanding world population, it will be necessary in the next 30 years 'to build houses, hospitals, ports, factories, bridges, and every other kind of facility in numbers that almost equal all the construction work done by the human race up to now.'[12] That work will require the expenditure of astronomical amounts of nonrenewable energy. When we stop to consider the worldwide energy problems we already face – the shortages, the mushrooming prices, the accumulating pollution and wastes – it should become more than obvious that the nation and the world

112

cannot meet the projected future energy needs, regardless of how hard we try. The reality is that we are running out of the 'available' stock of nonrenewable energy and moving dangerously close to an entropy watershed. While statistical studies vary, there is general agreement that the age of cheap nonrenewable energy is over.

A recent study, conducted under the auspices of MIT, and involving experts from industry, government, and academia from fifteen countries, concluded that the worldwide supply of oil 'will fail to meet increasing demands before the year 2000.'[13] According to this report, even if energy prices rise 50 per cent above current levels, it's likely that the world will experience an oil crisis between 1985 and 1995.[14] Another study undertaken by the powerful Trilateral Commission – an international organization whose membership includes some of the most powerful business and political leaders of the Western nations – varies only slightly in its forecast. The commission concludes that global demands for oil will exceed supply by the mid-1990s.[15] Writing in the *Bulletin of Atomic Scientists*, economist Emile Benoit of Columbia says that if international consumption of oil continues to grow at its present rate, existing reserves would be exhausted within twenty-five years. Even if new oil finds equal to four times the present reserves could be discovered (which is a highly inflated estimate that most experts consider unlikely), it would only buy an additional twenty-five years before the total exhaustion of all oil reserves.[16]

In his book *The Twenty-ninth Day*, ecologist Lester Brown computes that there are enough recoverable oil reserves to provide every American with approximately 500 barrels. When refined, says Brown, a barrel of oil yields about forty-two gallons of petrol. Consequently, if the average American drove 10,000 miles per year in a large car that gets around ten miles per gallon, he would use up his entire remaining share of the world's oil reserves in less than twelve years.[17]

Synfuels

With worldwide petroleum reserves becoming more scarce and more expensive to extract, process, and consume, efforts are being made to shift to other nonrenewable energy sources. Take, for example, synfuels – currently being touted as the saviour of modern industrial society. Actually, the very term *synfuel* is a misnomer. All so-called synthetic fuels are derived from nonrenewable resources.

American politicians are now heralding the United States as 'the Saudi Arabia of coal.' In reality, the abundance is not nearly what is forecast. Government bureaucrats insist that the country contains enough coal for 500 years of use. What they don't say is that at current coal consumption growth rates of 4·1 per cent annually, it is generally recognized that there are only enough coal reserves in the nation to provide for 135 years of energy.[18] While this still sounds rather impressive, as the country steps up coal production to offset declining availability of petroleum, the absolute time span will be greatly reduced to a few short generations. And as we push forward to extract the remaining coal from our land, we can expect that exorbitant government spending, higher inflation, and additional pollution will be the result.

Plans are now afoot to produce 1 million barrels of coal-based synfuel petrol by the year 1990. Even before the plan had left the drawing board, one White House official confidentially told a reporter that 'this has the potential for being one of the craziest schemes ever put forward by the Federal Government.'[19] It's not hard to see why.

First, the current technology for coal liquefaction requires over a ton of coal to be extracted, heated to a high temperature, and pressurized to squeeze out a mere three to four barrels of liquid oil. Not only will a tremendous amount of energy go into the refining process, but several hundred million tons of coal beyond that currently mined will have to

114

be extracted each year. In fact, it might not even be possible to mine the necessary coal because of the large amounts of water needed at each step of the synfuel process. According to one highly regarded survey, 'If all the coal mines, power stations and liquefaction or gasification plants now projected were to be built, they would require for their operations, exclusive of reclamation, between three and four times the *total* amount of water now used throughout the entire country' (emphasis in original).[20] To make the process even more dubious, most of the coal deposits and projected synfuel plants will be developed in the West, a part of the country where there is already precious little water.

Second, to process synfuels, an entire industrial infrastructure will have to be built from scratch. Currently, there is only one synfuel plant in the entire country. Located in Cattlesburg, Kentucky, the plant requires 250 tons of coal to make just 625 barrels of oil, barely enough to supply a single service station. To produce the 1 million barrels a day projected by the government, at least twenty synfuel plants, with a capacity of 50,000 barrels produced daily, will have to be constructed. Each of these facilities will be larger than most large petroleum refineries. The government estimates the cost at $2 billion per plant, but a Rand Corporation study predicts that, based on past cost estimates and overruns in other energy areas, the synfuel programme will cost at least 100 per cent more per plant. Total cost of the synfuel project, according to Rand – a whopping $100 billion, four times the amount invested in the decade-long Apollo programme to put a man on the moon. There are hidden costs, too. Massive pipeline systems must be built and the transportation system will have to be beefed up to allow for the shipment of billions of tons of additional coal. In all, the *New York Times* has estimated, the synfuel programme may cost $300 billion.[21]

Even this astronomical sum may prove on the low side. Because there does not exist in this country even a single synfuel plant of the size being advocated, there are sure to be unexpected design problems. But because the government is committed to a crash programme – to up synfuel production in one decade from 625 barrels daily to over a million –

building will have to proceed on a tight schedule. This 'can cause both higher cost and poor system performance.'[22] The Cattlesburg synfuel plant, for example, is costing nearly four times as much as predicted. Design and construction problems might be quite similar to those encountered in the nuclear power industry, where long shutdowns in the facilities have become commonplace.

Then there is the question of where to put the twenty massive plants under consideration. In the past decade, the country has managed to locate just one spot for the construction of a major oil refinery. This is largely due to the natural distaste residents of the community have for building huge, polluting plants in their area. Thus, while everyone seems to want more fuel, no one wants the plant that will produce the fuel in their backyard.

In fact, there are simply very few sites that are even remotely feasible as synfuel plant locations. A Department of Energy study concludes that only forty-one counties in the entire nation have coal reserves and available water supplies adequate to supply a massive synfuel plant. Some of these counties, such as in coal-rich Montana, Wyoming, Colorado, and North Dakota, have populations of less than 1,000. DOE estimates that a typical synfuel plant will bring in an additional 20,500 residents to each county. This will necessitate the construction of added community facilities – sewers, houses, streets, garbage collection, schools, etc. DOE projects that each community might find itself burdened by $70 million in additional costs.[23]

Aside from all of these problems, the environmental dangers of burning massive amounts of coal, in any form, make the synfuel project completely untenable. The monumental strip-mining of wilderness areas, and giant air-polluting synfuel factories, are just the beginning.

According to a National Academy of Sciences study, 'The primary limiting factor on energy production from fossil fuels over the next few centuries may turn out to be the climatic effects of the release of carbon dioxide.' Increased use of coal results in the emission of massive doses of carbon dioxide into the atmosphere. The carbon dioxide creates a warming or

116

'greenhouse' effect by blocking the radiation of heat into space. The NAS report says that carbon dioxide levels in the atmosphere could well double in the next seventy-five years, 'raising temperatures at the midlatitudes by 3° to 6° centigrade and that near the poles by about 9° to 12° centigrade.' The effect on the plant and animal life of the planet would be devastating. The entire ecological balance of the earth would be seriously jarred. Among other consequences, the polar ice caps would melt, raising ocean levels worldwide, causing the drowning of almost all major port cities around the globe. Dramatic changes in world temperature would result in the wholesale extinction of much of our existing plant and animal life. The speed of the change alone – seventy-five years or less – would eliminate any possibility of evolutionary adaptation. (It would take millions of years for most plant and animal species to adjust genetically to such quantitative changes in the earth's temperature.)[24]

The NAS is not alone in its concerns. Many reports over the past several years have come to the same conclusion. World Watch Institute argues that if current emission rates of carbon dioxide continue for several more decades, the resulting world-wide temperature increase will produce climatic changes 'perhaps on a scale approaching that with which the Ice Age came and went.'[25]

Synfuels seriously magnify this problem. The Council on Environmental Quality says that synfuel use will double the speedup of the greenhouse effect because synfuels produce more CO_2 than regular fuel. For the same amount of heat, and work, synfuels put out 1·4 times as much carbon dioxide as coal, 1·7 times as much as oil, and 2·3 times as much as natural gas. A national commitment to synfuels, CEQ warns, could result in the polar ice caps melting by the end of the century.[26]

The laws of thermodynamics tell us that it takes energy to transform energy. Net energy is the total energy that remains once we have subtracted the energy required to manufacture it. Viewed in these terms, synfuels are in no sense energy efficient. A 50,000-barrel-a-day synfuel plant will consume up to three times the coal used at a 1,000-megawatt coal-burning plant. In addition, if we take into consideration the amount of energy required to mine the coal, produce steel to build the behemoth plants, and construct pipelines and new transport systems, we find that while oil yields 50 BTUs for every BTU that is used to produce it, coal synfuels yield just 17 BTUs and shale oil only 6·5 BTUs for every BTU used to produce them.

Other aspects of the synfuel programme share similar problems. Take, for example, shale oil. It takes a ton and a half of oil shale to produce just one barrel of oil, and two barrels of water are required in the manufacturing process. Shale oil also produces highly toxic fumes which have to be disposed of. As for tar sands, some 4,400 pounds of sand must be mined and heated to produce just a single barrel of oil. Again, the net energy yield is very small due to the large amount of energy required to transform one form of energy (shale and tar sands) into another (liquid oil). Even with the most advanced technological extraction processes, it is estimated that shale could provide only about 2 per cent of the current energy consumption needs of the United States.[27]

Nuclear Fission

Until recently, the hope for an energy alternative rested with nuclear power. Now, that hope is fading fast. Even before the near meltdown of the nuclear core at Three Mile Island in March of 1979, serious problems inherent to the nuclear industry began to spell a dismal future for 'the peaceful atom.'

Extraordinarily high production costs have combined with severe health and safety concerns to greatly reduce the number of nuclear power plants being built. Projections of nuclear energy growth are now less than one-third of what they once were in many countries. In the United States there were 36 new nuclear power plants ordered in 1973 and 27 in 1974. In 1975 the orders dropped to 4, in 1976 to 2, and in 1977 back up to only 4.[28] Aside from the fact that a nuclear power plant can cost up to $2 billion to construct, hidden costs deflate the 'cheap energy' myth of atomic energy. A congressional report made public in 1978 states:

Contrary to widespread belief, nuclear power is no longer a cheap energy source. In fact, when the still unknown costs of radioactive waste and spent nuclear fuel management, decommissioning and perpetual care are finally included on the rate base, *nuclear power may prove to be much more expensive* than conventional energy alternatives such as coal [emphasis in original].[29]

In addition to the cost factor, nuclear power generates many social and health problems for which there is simply no technical solution. Mining the uranium needed for the reaction can not only lead to cancer and other diseases among the miners, but can cause serious health side effects in communities located near uranium mines. Already, 100 million tons of uranium tailings (waste ore that remains once the uranium is removed) have accumulated in the southwestern states. This waste has a radioactive half-life of 80,000 years. In Colorado, where tailings have been used by construction firms as

119

building materials in schools and homes, doctors have noted an increase in congenital birth defects among children whose parents live or work in buildings made of tailings.[30]

The reactors themselves are not at all safe, either. The Three Mile Island accident was but the most serious of scores of breakdowns and radiation releases experienced by the nuclear industry. Every reactor in the country is constantly leaking small amounts of radioactive material into the environment. The industry, of course, is always quick to point out that the radiation discharge is below officially acceptable limits. What is never said, however, is that the medical evidence suggests that all radiation, in however minute an amount, is potentially dangerous. Every dose of radiation is an overdose. It takes just one radioactive particle invading one cell to cause cancer or genetic mutations. The process, however, can be an insidiously slow one, taking perhaps two decades from the time of exposure to the onset of the disease. Because of that, we may be unleashing a future epidemic with today's nuclear power plants. The Union of Concerned Scientists has estimated that by the year 2000, close to 15,000 Americans will have died as a direct result of minor reactor accidents and leaks. Should there be a full 'China syndrome,' UCS predicts that 100,000 might die, and thousands of square miles of land would be contaminated for many years to come.[31]

An even more serious problem is that each nuclear reactor produces between 400 and 500 pounds of plutonium yearly. Plutonium is the basic raw material used in constructing nuclear bombs. At the current rate, every reactor in the country annually generates enough plutonium to manufacture up to forty atomic weapons. Within two decades, there will be enough fissionable material in international transit to make 20,000 nuclear bombs. Guaranteeing the safety of this material is an impossibility. Already, 700 pounds of plutonium is missing from reactors and storage sites around the country.[32] Given that all the technical knowledge needed to build an A-bomb is already available on library shelves across the country, the continued production of plutonium is simply inviting someone to make and use their own nuclear device.

In a study conducted by the Office of Technology Assessment, *The Effects of Nuclear War*, it was found that a relatively small terrorist device could completely destroy several blocks of a high-rise downtown area, sending out over a thousand times the amount of radiation considered allowable for human exposure, and causing deadly fallout in the suburbs. In fact, a bomb isn't even necessary to cause incredible death and mayhem. If plutonium were dispersed in the open air over a city, an area of forty square miles would be contaminated for a period of 100,000 years.[33]

Then, too, there is the unresolvable problem of how to dispose of nuclear waste. As hard as it may be to believe, with all of the attention placed on nuclear research and development and after spending billions of dollars to erect existing plants, the scientific community, the energy companies, and the government have not yet figured out how to get rid of the radioactive waste. Says Harvey Brooks of Harvard University, who heads the National Academy of Sciences committee on the question of nuclear waste disposal: 'I would predict that should nuclear energy ultimately prove to be socially unacceptable, it will be primarily because of the public's perception of the waste disposal problem.'[34]

By the end of 1976, 3,000 metric tons of spent fuel rods lay in nuclear pools in the United States. By 1983, the amount is expected to increase to 13,000 tons. Solid wastes – contaminated clothes and equipment – are another huge problem. Already, there are 13 million cubic feet of radioactive solid wastes, containing 2,200 pounds of plutonium. And the nuclear industry estimates that by the year 2000, there will be a total of 152 million gallons of high-level liquid waste as well.[35] While there are many plans centring on how to 'safely' bury this massive amount of waste, none has proven effective simply because there is just no way to guarantee that a lethal substance can be stored for thousands of years. After all, our nation has only existed for 200 years. Human civilization is but a few thousand years old. Imagine having the audacity to think we can devise a programme to store lethal radioactive materials for a period of time that is longer than all of human culture to date.

The long-term question aside, the fact is that the industry hasn't even succeeded in finding adequate storage measures for the 1980s. Even with today's relatively small amount of nuclear waste, there are constant reports of leakage and accidents at dumping sites. Radiation leaks have been discovered at the US Government nuclear reservation in Richland, Washington. Over 500,000 gallons of liquid radioactive wastes have leaked from tanks stored at the facility. In June 1978, the state of Kentucky closed down its nuclear site at Maxey Flats in the wake of an EPA study showing that 'radioactive particles were migrating offsite.' Similar leakages have been reported at burial sites in Oak Ridge, Tennessee; Ocean City, Maryland; and near San Francisco, California.[36]

Even if US nuclear power continues to level off, it will be necessary to find new burial sites every two to three years after the turn of the century to accommodate all of the waste. This in turn will necessitate strict monitoring and armed guards around the clock on each site for up to 250,000 years to insure against leakage into the biosphere: that is the average time needed for some of the radioactive wastes to become harmless.

Nuclear Fusion

While nuclear energy generates problems that appear to be insurmountable, the nuclear genie dies hard. Twenty years ago, Americans were promised that fission power would usher in an age of 'limitless, clean and too cheap to meter' energy. Today, even as this myth has been dealt a severe blow by accidents like the one that occurred at Three Mile Island, a new nuclear promise is being touted: the promise of fusion power. The claims of its proponents are eerily reminiscent of those made by fission advocates two decades ago.

Technically, fusion power is the reverse of fission. Instead of splitting apart one nucleus, as in fission, fusion slams together (fuses) two nuclei from different atoms. Fusion energy is nothing new; it is constantly taking place in the sun, releasing the life-nurturing energy that has bombarded our planet for billions of years. In the 1950s, humans learned how to unleash a fusion reaction through the hydrogen bomb. Scientists now hope to discover a method to harness the immense energy that can be generated in such an explosion by containing the reaction within a fusion power plant.

Proponents of fusion energy argue that the process is more efficient than fission, generates far less radioactive waste, and might one day rely for its fuel upon hydrogen, which could conceivably be taken in virtually unlimited quantities from the ocean. In a sense, fusion energy is the modern-day equivalent of the perpetual motion machine. But just as the Entropy Law precludes the workings of a perpetual motion machine, it sets severe restrictions upon the viability of fusion energy.

To begin with, no one can say for certain whether a contained fusion reaction can even be sustained. To be commercially useful, a reactor must fuse 100,000 billion hydrogen nuclei per second within each cubic centimetre of the reactor core. To date, a contained fusion reaction has been sustained for only a fraction of a second. Unless the reaction can be prolonged, the fusing process will require immeasurably more energy than it will produce. The most optimistic assessments speculate that it will be at least 2025 before fusion will be able to produce any commercial power, hardly soon enough to meet the energy crunch that now besets the world.[37]

Second, while there are several kinds of fusion technologies, the type being explored now is called a deuterium-tritium reaction, because the process fuses together molecules of these two elements. Tritium is derived from lithium, a nonrenewable resource that is almost as scarce as uranium. Thus, fusion energy is not limitless; it can only be sustained as long as the lithium stock of the world remains. A fusion plant will also require huge amounts of other nonrenewable energy-intensive resources which are already becoming increasingly scarce, such as niobium and vanadium. Each 1,000-megawatt plant will require the mining of an additional 2·8 million pounds of copper, a metal that is also becoming relatively scarce.[38]

Third, the 'clean' nature of a fusion reactor is a peculiar kind of cleanliness. Miners will still be affected by the extraction of lithium, just as uranium miners are now. Fusion reactors can hardly be said to be waste free, either. A large fusion plant might produce as much as 250 tons of radioactive garbage yearly.[39] The same containment problems that plague fission reactors would remain.

Further, there are tremendous technical and maintenance problems associated with any foreseeable fusion reactor design. One of the reasons scientists are experimenting with deuterium-tritium fusion is because it can operate at temperatures of 100 million degrees centigrade. (The hydrogen-boron reactor, which could be fuelled with sea water, has a reaction temperature of 3 billion degrees.) These rather mind-numbing figures take on special meaning when it is realized that, at this point, we know of no materials that can withstand such sustained heat and tremendous radiation. Dr Bowen R. Leonard, Jr., senior scientist at Battelle Pacific Northwest Laboratory, says that the heat and radiation generated by the reaction might make fusion power prohibitively expensive. 'Radiation at that level destroys the strength of steel or other structural materials ... making it dangerously brittle very fast. Parts would have to be replaced constantly and shutdowns would be frequent.'[40] Some parts of the structure – such as walls nearest the reaction – might need to be changed every year, but because of the intensely radioactive nature of the reaction, no human could safely do the maintenance work. A new generation of industrial-maintenance robots will have to be designed at tremendous cost. No one even knows how long a plant will last, but estimates hover around the twenty-five-year mark. Once the plant has been stressed beyond its operational capacity, it will have to be dismantled, transported, and buried.

In addition to all of these technical and resource questions, says Amory Lovins, a physicist and proponent of solar power, fusion energy represents 'a way of doing something we don't want to do: that is build a complex, costly, slow-to-deploy, centralized, high-technology way to make electricity.' All nuclear technologies, he believes, are the equivalent of using a chain saw to cut butter.[41]

Minerals

Energy – whether in the form of oil, coal, uranium, or solar – cannot be viewed in isolation. If we want to extract energy from our environment, it must be done by using nonrenewable resources in the form of drilling rigs, tractors, and plants. And if we want the energy to perform work, it can only do so in conjunction with more nonrenewable resources such as those invested in machines and factories. Because of this, energy resource depletion is only part of the story of the physical limits we are now experiencing on our planet. The earth is fast running out of almost every major nonrenewable mineral necessary for the maintenance and growth of highly industrialized economies. Each year the US economy alone uses nearly '40,000 pounds of new mineral supplies per person for our power plants, transportation, schools, machine tools, homes, bridges, medical uses and heavy equipment.'[42]

America is chiefly responsible for gobbling up the remaining stock of the earth's precious minerals. According to the US Department of the Interior, the US economy produces or imports 27 per cent of the world's bauxite production, 18 per cent of the world's iron ore production, and 28 per cent of the world's nickel.[43] In order for the rest of the world to reach a par with the American standard of living, it would have to consume up to 200 times the present output of many of the earth's nonrenewable minerals (this assumes a doubling of world population between now and the early part of the twenty-first century). While catching up with the US standard of living is the goal of most developing nations, it is obviously a pipe dream.[44]

Many experts predict that within seventy-five years or less (at current consumption rates) the planet's economies will have 'exhausted presently known recoverable reserves of perhaps half the world's now useful metals.'[45] Dr Preston Cloud, a geologist at the US Geological Survey, is one of those experts. Testifying before the Joint Economic Committee of the Congress in 1978, Cloud said that some of the minerals on the endangered resources list by early in the next century include copper, gold, antimony, bismuth, and molybdenum.[46] Domestically, by the year 2050 the United States will have run out of extractable quantities of tin, commercial asbestos, columbium, fluorite, sheet mica, high-grade phosphorus, strontium, mercury, chromium, and nickel.[47]

Increased reliance on foreign imports of most key minerals, combined with intense worldwide competition for the remaining scarce reserves, will raise prices and the bargaining leverage of the mineral exporting countries – just as was the case with oil for the OPEC nations.

The flow of nonrenewable resources through society also affects the consumption of renewable resources. While it is true that forests and fish are living organisms that create more of their own kind, the annual consumption of these resources appears to be increasing at a faster rate than they can be replenished. In effect, the high-entropy economic system 'hot-wires' the renewable resources to the point where they become, for all practical purposes, nonrenewables themselves. Before the advent of the fossil fuel era, humanity relied almost exclusively on forests, fisheries, grasslands, and croplands for its energy flow. Now, however, evidence suggests that the productivity of each of these systems has peaked and is declining. Global forest productivity has steadily declined since 1967. Fisheries peaked in 1970, and now many traditional fishing areas of the ocean have essentially been 'fished out.' Cropland productivity, as measured by kilograms of cereals per capita per year, peaked in 1976. As for grasslands per capita output of wool, mutton, and beef (all dependent on grazing) have all tailed off.[48]

Despite the overwhelming statistical evidence presented in United Nations, congressional, and academic studies, reports, and hearings, there are still a few souls who cling to the theory that at existing growth rates there are enough nonrenewable resources to provide for all the world's population forever – or at least for a good long time into the future. Their underlying assumptions, however, are without merit.

For example, it is often remarked that the entire planet is composed of minerals. What is overlooked is that only a tiny fraction of that amount is usable or potentially extractable. Assume, just for the sake of argument, that the entire weight of the earth was potentially convertible to productive energy – which would leave us all walking on thin air. At a current 3 per cent growth rate in the use of ten leading minerals, we would literally mine the equivalent of the entire world's weight within several hundred years. That's not a very long time when one stops to realize that human beings have been on earth for over $3\frac{1}{2}$ million years and that the earth itself has existed for 4 billion additional years.

Others argue that manganese nodules mined from the seabed could provide us with a source of nonferrous metals. According to some experts, this source could provide 'copper equal to a quarter of current output, nickel equal to three times current output, and manganese equal to six times current output.' These same experts believe it's possible to quadruple these figures sometime in the future. Again, these figures appear impressive on the surface until they are placed within the context of exponential growth. At current rates of consumption increase, demand for copper will be 90 times the current level in just a hundred years, nickel 28 times the current level, and manganese 17 times the present level, virtually wiping out whatever short-term advantage measured in years or decades that might accrue from these additional deposits.[49]

Substitution, Recycling, and Conservation

There are those who continue to believe that existing reserves of nonrenewable minerals can be maintained indefinitely by either replacing more-scarce minerals with less-scarce ones or by efficient recycling of existing mineral use. As to substitution, since most major metals are fast diminishing in supply, there is relatively little advantage in substituting one for another in the production process. As William Ophuls points out, 'Substitutes (like aluminium for copper) are on the whole less efficient than the material they substitute for, and more energy is therefore required to perform a given function.'[50] Then too, some minerals because of their unique properties are simply irreplaceable.

Recycling is often latched onto as the answer to mineral resource depletion. Recycling already provides about half of the annual demand for antimony; one-third the demand for iron, lead, and nickel; and one-fourth our need for mercury, silver, gold, and platinum. However, it should not be forgotten that recycling also conforms to the second law of thermodynamics. Every time a mineral is recycled, some of it is inevitably, and irreversibly, lost. As already mentioned, recycling efficiency today averages around 30 per cent for most used metals. Recycling also creates additional pollution and requires ever greater amounts of energy input 'to collect, transport and transform' the scattered material. Like metal substitution, recycling, within the context of existing exponential growth rates in mineral use, buys only a small, almost irrelevant period of extra time – a few decades, maybe fifty years at most. While more efficient recycling is going to be essential in the future, data indicate that little more than 1 per cent of our total mineral needs can be met in the foreseeable future through recycling.[51]

Conservation, like recycling, is undeniably of value; but again, like recycling, it can only be a partial solution. It is often pointed out that nations in Europe with standards of living equivalent to our own consume only half the energy per capita as the United States. (A fitting rejoinder to this, of course, is that even if the United States used only half of the current energy flow, the essential problems of exponential growth, resource depletion, pollution, and so forth would remain. Europe, after all, is hardly an environmental paradise.) While conservation is absolutely essential, the fact is, any current conservation proposals will be extremely limited in scope because they have to be implemented within the existing high-energy infrastructure. Any attempt to extend conservation efforts beyond the narrow confines dictated by the current infrastructure can only lead to serious dislocations at various points along the energy flow line.

The current federal effort to cut down on electricity consumption by setting thermostats on the air conditioning systems higher, show, in microcosm, the kinds of disorders that are caused by attempts at conservation in a system designed for high-energy flow-through. America's addiction to air conditioning is legendary. The first air cooler was installed in 1922 in a movie theatre. Today, Americans consume more electricity for air conditioning during the three summer months alone than does the entire population of the People's Republic of China to meet all of their annual electrical needs. And China has four times the number of people![52]

Begun as a gimmick and a convenience, air conditioning has become an essential part of our social system. During the past two decades, when energy was moving through the flow line at unprecedented rates, buildings in every part of the country – from the World Trade Center in New York to Holiday Inns in California – were designed with windows that do not open. At the time, everyone assumed that there would be plenty of electricity to guarantee yearly air-conditioned comfort in a completely closed environment. Yet, if thermostats are to be set higher, occupants of the buildings need fresh air and occasional breezes. But the only way that can be accomplished is by changing millions of windows in hundreds of thousands of structures – clearly a massive use of energy, resources, time, and people.

There are other implications resulting from this simple attempt to conserve electricity. Heating and cooling systems in many new buildings were specifically designed to operate most efficiently at temperatures below the limit the government has set. In order to reach the higher temperatures and conform to the regulations, some building operators are finding that they actually have to *heat* the air ducts to raise the temperature.

The list goes on and on. Numerous processes we take for granted have been specifically designed for an air-conditioned world. For example, computers require low humidity and cool air to function properly; if the air conditioning is turned off, they cannot work. Banks have found that turning up the thermostat wreaks havoc with the glue on envelopes, which becomes sticky when the humidity rises. Many customers in restaurants have quickly learned that their favourite eateries were designed in such a way that the heat emanating from the kitchen ovens can only be controlled by air conditioning.

Altering thermostat settings even has biological and psychological effects. In a sense, many of us have adapted to air conditioning so thoroughly that summer temperatures and humidity which humanity has easily endured for 99 per cent of its history now seem intolerable. One study shows that workers in offices with less air conditioning actually produce significantly less than those in cooled comfort. Once again, the Entropy Law makes itself felt.

This is but one of tens of thousands of examples we could point to. None of this is to deny the absolute necessity of conserving as much as possible. But conservation works to the extent that the system is designed to accept a lower energy through-put. Industrial, urbanized society is specifically designed for just the opposite purpose – to maximize energy flow. Given this basic fact of modern life, conservation measures within the existing high-energy infrastructure can only serve as a palliative.

PART FIVE

Entropy and the Industrial Age

Economics

The industrialized nations, and the United States in particular, are coming up against an entropy watershed. After 400 years, the world is now running out of the nonrenewable resource base that provided the industrial era with a massive flow-through of stored solar energy. At every stage of the energy flow line, disorder is mounting, and the technological and institutional transformers are becoming more complex, more concentrated, more specialized, and more vulnerable to breakdown.

One doesn't need to be an economist to understand the process. Since we all survive by transforming, exchanging, and discarding energy in all of its many forms, we experience first-hand the tremendous dislocations in the energy flow line as the society moves closer and closer to an entropy watershed. Nowhere is the process more apparent than in dealing with the ravages of inflation. On every major opinion poll taken over the past five years, the American people have listed inflation as their number-one concern. Says financial columnist Sylvia Porter:

Unless the battle against inflation is fought with more courage and originality than it has been up to now, we, the people, are moving closer and closer to taking our destiny into our own hands. The rage of the 1960s then may be child's play against the fury of the 1980s.[1]

What Porter and other columnists fail to completely realize is that today's inflation is tied directly to the depletion of our nonrenewable energy base. As it becomes more costly to extract less easily exploitable supplies of available energy from the environment, the costs associated with all of the transforming, exchanging, and discarding processes all along the energy flow line continue to rise. As a result, prices continue to rise for both the producer and the consumer. The accumulating disorder resulting from past flow-through adds

135

additional economic, social, and political costs, further increasing prices for producers and consumers. The inflation spirals faster and faster as the energy environment nears depletion. The reason again is simple: it takes more and more money to pay for more expensive, complex technology to extract and process the remaining energy and more money to pay for controlling or managing all of the disorder resulting from the dissipation of energy in the flow-through process.

According to Dr Barry Commoner, one of the nation's top energy experts, all of the basic energy sources that we rely on suffer from the same flaw:

Because they are either non-renewable, or over burdened with unnecessarily complex technology – or both – they demand progressively larger investment of capital, become increasingly costly to produce, and – in the free market of the private enterprise system – higher in price.[2]

Commoner then points to statistics which provide irrefutable evidence of how the Entropy Law effects the whole process. For every dollar invested in energy production in 1960, 2,250,000 BTUs of energy were produced. In 1970, says Commoner, every dollar invested was only producing 2,168,000 BTUs of energy. Just three years later, in 1973, the figure had dropped to 1,845,000 BTUs for each dollar invested. In just thirteen years there had been a *decrease* of 18 per cent 'in the productivity of capital in energy production.'[3] (The data, by the way, were adjusted in terms of 1973 dollars to eliminate the effects of inflation.)

Because it costs more to extract less-available energy out of the environment, the amount of money that has to be diverted away from the rest of the flow line to pay for new capital for the energy industry continues to climb. The energy industry will need to raise over $900 billion to finance its operations over the next decade. Over half of that amount, however, will have to be raised externally, because these companies do not have sufficient retained earnings to finance their own needs internally.[4] This means that monies normally invested in other areas of the economy will have to be diverted to maintain the energy industry. In the early 1970s, nearly 24 per cent of the total capital invested in US industry went to support energy production. By 1985, it is estimated that over one-third of all investment capital will have to be diverted to the production of energy just to maintain projected needs.[5]

As more and more money is diverted into energy production, the energy transformers – both the machinery and the institutions – become more concentrated, complex, and powerful. Today America's energy institutions own $181 billion in assets or '29% of the assets (and sales) of the 500 largest corporations in the US'! So large are energy companies such as Mobil, Exxon, and Texaco, that twenty of them now account for 18 per cent of our total gross national product.[6] With the costs of a new petroleum refinery at $500 million, and the cost of a nuclear power plant at between $1 and $2 billion, only these corporate giants can afford to stay in the energy game.[7]

Energy, of course, is the basis for all economic activity. Therefore, as the costs go up at the source, they are passed along in terms of higher prices at each succeeding step in the energy flow line. Eventually the individual consumer pays the bill in terms of inflation.

The Exploratory Project for Economic Alternatives, a Washington think tank, recently undertook a detailed study of the basic causes of inflation. It concluded, in its final report, that in the four basic consumer necessities – energy, food, housing, and health care – the rising prices were tied to the increased costs associated with the transforming and exchanging of energy. While this seems obvious enough, most establishment economic thinking continues to centre on secondary effects like wages or fiscal and monetary policy.

For 80 per cent of American families, the four basic necessity areas comprise over 70 per cent of their consumption budgets. The study isolated each of the necessity areas and then traced the cause of the inflation back to its source – the depletion of the nonrenewable energy sources, and the increased technological, structural, and institutional costs associated with the continued maintenance of the energy flow-through. For example, energy alone – petrol, electricity, fuel, oil, and coal – accounts for nearly 12 per cent of the average household budget. The study found that energy inflation has 'robbed ordinary Americans of 1% of household purchasing power per year.'[8]

In the case of food, which accounts for approximately 28 per cent of the average family budget, the inflation rate has been moving at 8 per cent or more per year. The report traced the inflation trend back to the higher energy costs associated with the farming, processing, transporting, packaging, and marketing of agricultural products, as well as the increased worldwide population demand for American foodstuffs.[9] In the area of housing and health care, again, the higher cost of nonrenewable energy was at the source of the inflationary spiral. That's because all economic activity is traceable back to the prevailing energy base.

Inflation, then, is ultimately a measure of the entropy state of the environment. The closer the entropy of the environment moves to a maximum, the more costly everything in the energy flow line becomes. As already shown, the costs associated with transforming energy rise as the sources of energy become more difficult to locate, extract, and process. The cost of exchanging energy between institutions, sectors, groups, and individuals also rises to reflect the increasing costs associated with the extraction and processing.

We've already seen how the consumer is ultimately affected by the higher costs in terms of being charged higher prices for basic necessities. The wage earner is also affected. That is, while wages rise, real purchasing power fails to keep up with the rise in the cost of living. 'Average weekly earnings in 1976 in constant dollars (consumer price index deflated) were still below their 1971 levels.'[10] The growing gap between wages and real purchasing power is the money that is diverted away from the labour bill to pay for the increased costs of maintaining the nonrenewable energy flow. It works this way: as the costs go up at the beginning of the energy flow line, they are passed on to every other economic institution down the line as well. To compensate for the increased costs, all of the economic institutions along the entire flow line from extraction through to retail sales attempt to reduce their wage component in order to maintain existing profit levels. The result is less 'real' wages and less purchasing power. Less purchasing power really means that consumers are increasingly unable to meet their energy needs: food, clothing, health care, etc. In other words, as mentioned earlier, the energy flowing through the human system begins to slow down, as more and more energy (or money) is diverted to the maintenance of all the economic institutions and the machinery responsible for the energy flow itself.

While the consumer suffers from higher prices and the worker from lower real wages, the taxpayer suffers from the increased costs associated with the dissipated wastes and disorders that build up along the flow line. It is the taxpayer who has to pay the lion's share of cleaning up and disposing of the massive wastes generated by the flow of energy through the system. According to the annual report of the President's Council on Environmental Quality, the taxpayers shelled out nearly $16 billion in 1977 to pay for pollution control, and the costs are expected to escalate at 20 per cent per year. The council estimates that the overall costs of pollution control over the next ten years will exceed $361 billion, much of it paid for by government tax dollars.[11]

The taxpayer also ends up paying for the economic and social disorders that arise as a result of the way the energy flow line is set up. For example, certain individuals, groups, and classes are located at the periphery of the transforming and exchange processes because of the way the system allocates jobs and distribution of income. As the entropy of the environment increases, and the costs all along the flow line escalate, this sector of the population is the first to feel the economic crunch. As more and more people in the poorer classes are thrown off the flow line altogether to compensate for the tightening economic condition, the government must step in and provide for their energy needs in terms of welfare and other benefits. Unemployment, after all, is the other side of the entropy process. The faster the energy is depleted, the more people become either unemployed or underemployed. The government institutions at all levels from local to federal have to enlarge their own involvement in order to provide relief for these front-line victims of the tightening energy crisis.

139

The government also has to enlarge its functions in other areas that are directly affected by the increase in unemployment and poverty such as crime control and public health expenditures. In addition, still more money in the form of taxes has to be diverted away from the flow line in order to pay for the increased costs of maintaining these public bureaucracies. Today, 16 per cent of the American work force is employed by a public institution or a governmental agency of some sort.[12] These government institutions continue to grow as they are forced to deal with and contain the mounting economic and social disorders arising along the flow line. Like the economic institutions, however, the government agencies end up using more and more money just for their own maintenance, thus increasing the tax burden and further decreasing the flow-through of energy for human consumption. This vicious process of energy diversion away from the people and toward the maintenance and enlargement of both the economic and governmental bureaucracies proceeds faster and faster until the entire social mechanism crashes headlong into an entropy watershed.

It should be clear by now that classical economic theory cannot solve the growing crisis facing the world's economies. There is no room in either socialist or capitalist economic analysis for the Entropy Law. Yet, the second law is the supreme governing principle of all economic activity. The failure to recognize this ultimate truth and to reorient all economic policy around it is shortening the road to economic and ecological disaster for the planet.

Today, as 200 years ago when Adam Smith first laid out the principles of modern economic theory, both socialist and capitalist nations model their economic assumptions along the lines of the classical mechanical doctrine. Behind every economic policy statement looms the shadows of Newton, Descartes, Bacon, Locke, and Smith.

Capitalist economists continue to view the economic system as the mechanical process in which supply and demand functions are continually readjusting to each other in forward and backward motion like the swing of a pendulum.

140

Pick up any introductory economics textbook and it will tell you that economics is nothing more than the give and take of supply and demand curves. When consumer demand for a commodity or a service goes up, the sellers will raise their price accordingly to take advantage of the situation. When the price becomes too high, demand will slacken off or move to some other commodity or service, forcing the sellers to lower the price to the point where demand is rekindled and so on. Many qualifications and refinements have been added over the years, but the basic concept of the market mechanism of supply and demand still remains at the centre of all classical economic thinking.

While socialist economists reject the market mechanism, they agree with the capitalist economists that the overall economic environment is never depleted. As to where the new supply is supposed to come from, both capitalist and socialist economists assume that new technology can always find a way to locate and exploit previously untapped resources. The resource base itself is considered inexhaustible.

According to capitalist and socialist theory, economic activity turns waste into value. Remember Locke's belief that everything in nature is to be considered waste until human labour is added to it, transforming it to something of value that can be exchanged and consumed in society. By turning the first and second laws upside down, modern economic theory has completely misinterpreted the entire basis of all economic activity. Again, the first law says that all matter-energy is fixed, that it can neither be created nor destroyed but only transformed. The second law, in turn, says that it can only be transformed in one direction, from available to unavailable, or from usable to unusable. Whenever energy is extracted from the environment and processed through society, part of it becomes dissipated or wasted at every stage, until all of it, including that which is made into products, ends up in one form or another as waste at the end of the line.

Most economists simply can't buy this simple truth. They are wedded to the idea that human labour added to nature's resources creates greater value, not less. Because machine capital is ultimately viewed as past human labour mixed with resources, it too is considered to be creating economic value. They can't get it into their heads that machines and people can't create anything. They can only transform the existing available energy supply from a usable to a wasted state, providing only 'temporary utility' along the way.

Economists steadfastly cling to the idea that human labour and machinery create only value, because they believe in the paradigm of permanent and unlimited material progress. But we know from the second law that every time human energy or mechanical energy or any other form of energy is expended to make something of value, it is done at the expense of creating even greater disorder and waste in the overall environment. We also know that even the things of value that we make eventually end up as waste or dissipated energy. Thus, there is no such thing as 'material' progress in the sense of accumulating a 'permanent' store of usable goods, for everything we make in the world eventually ends up as dust in the wind.

The implications here are extraordinary. Consider for a moment the concept of productivity. Capitalist and socialist systems define productivity in terms of speed per unit of output. A premium is placed on performing a given task as fast as possible. A more appropriate thermodynamic measure of productivity would emphasize the entropy produced per unit of output as opposed to speed per unit of output. A study was done several years ago on how much energy was required to make an automobile. The study concluded that many times more energy was actually used than was necessary to make the automobile. Why was all the additional energy expended? To get the automobile off the assembly line faster. The greater the emphasis on speed of conversion, the more energy is used up beyond what is essential in making the product. Much of the energy wasted in modern industrial economies is the price we pay for speed.

Interestingly enough, anyone who has ever been caught with a nearly empty petrol tank while driving along a lonely highway, has understood the difference between defining productivity by speed per unit of output vs. entropy produced per unit of output. Faced with the prospect of running out of petrol and not knowing how far up the road the next petrol station might be, the driver has two choices available to him. He can accelerate in order to try and reach the petrol station faster. Or he can drive slowly. It is not surprising that many of us, when faced with this situation, generally react by accelerating, believing that the increased speed will somehow enhance our chances of getting to that filling station. In truth, the opposite is the case. By using the petrol more judiciously, a further distance can be travelled. It takes longer, but the time lost is made up by the energy saved that can be used to travel a longer distance. In terms of thermodynamic efficiency, then, productivity is a measure of entropy produced per unit of output, not speed per unit of output.

The Entropy Law also tells us that every time we increase the rate of energy expended by either human or machine labour, the decrease in entropy or enhanced value of the product results in an even larger increase in disorder somewhere else in the overall environment. Therefore as long as productivity is measured in terms of speed per unit of output, more energy will be used than necessary in converting resources into economic utilities and this increased energy flow-through will result in greater disorder or entropy build up which ultimately has to be paid for by the society. 'Haste makes waste' is an age old adage that reflects an intuitive understanding of the Entropy Law at work.

As long as there was an abundant supply of fossil fuels and the particular metals used to fashion and maintain the industrial mode, it seemed logical to define productivity in terms of speed per unit of output. Now that the existing matter-energy base is becoming depleted, and the entropy from past economic activity is accumulating at a rate beyond the system's ability to absorb, a major reformulation of the notion of productivity will need to be made by the economists in order to adjust to the requisites of thermodynamic efficiency in the economic process of production and consumption.

The economics profession has still not understood that 'the entropy law is the basic physical coordinate of scarcity.'[13] Nowhere is this more obvious than in discussions about 'balancing budgets.' While it is generally recognized that a society cannot continue to consume faster than it produces, the economists remain ignorant of the fact that the ultimate balancing of budgets is not within society, but between society and nature. This failure to understand the larger environmental context within which economic activity takes place is the key to understanding why economic theory has been unable to address the problem of deficits. Approximating a balanced budget requires that society not consume faster than nature can produce. Ecosystems operate as near to a steady state as possible (the laws of thermodynamics tell us that a perfect steady state is impossible to attain). The entire conversion process from low entropy to high entropy is maintained at a speed commensurate with the system's ability to manage a relative balance between production and consumption. Wastes are generated, absorbed and recycled for reuse maintaining a balanced ecological cycle. While 100% recycling is thermodynamically impossible, natural ecosystems come as close as possible to the ideal state of a balanced budget between production and consumption.

Economic activity is merely human intervention into the ecological cycle, borrowing low entropy inputs, converting them into temporary utilities and eventually discarding them back into the ecological cycle in the form of high entropy wastes. If society borrows low entropy matter and energy, converting it into utilities, then waste, at a speed far exceeding the conversion process in nature itself, the deficit escalates. Wastes are dumped back into the system faster than they can be absorbed and recycled, creating mounting disorder in the environment and escalating external costs for society. At the same time, available matter and energy is being depleted faster than nature can reproduce, causing increased scarcity in nature and mounting supply costs for society. Of course, it should be recognized that even if society attempted to balance its budget with nature, learning to consume no faster than nature produced, it is impossible to eliminate a growing deficit. The most that can be done is to slow down the rate. That's because any use of nonrenewable resources like oil and natural gas automatically increases the deficit. For all practical purposes these nonrenewable resources represent a fixed store of ecological capital that can only be used once.

By minimizing the use of nonrenewable resources and by using up renewable resources only as fast as they can be replenished without inflicting severe damage to the ecological cycle, it is possible to minimize the deficit between consumption in society and production in nature.

Closely related to the misunderstanding about the nature of balanced budgets and deficits is the problem of money and debt. Over the years a few scholars like Frederick Soddy and Herman Daly have attempted to point out the obvious contradiction that exists between the social conventions of money and debt on the one hand, and the entropic flow of nature on the other, but their critique has been ignored within the economics profession. Money, for example, is a form of national debt. It represents a lien against the total physical wealth of the community which an individual is free to exchange for actual physical wealth sometime in the future. The problem that economists completely ignore is that the generation of physical wealth by the community is not inexhaustible. The laws of thermodynamics set ultimate limits to the amount of physical wealth that can be generated. However, there is no limit to how much money can be produced and put into circulation. The problem becomes apparent with the introduction of debt and compound interest. As Nobel chemist Frederick Soddy pointed out over 50 years ago:

Debts are subject to the laws of mathematics rather than physics. Unlike wealth, which is subject to the laws of thermodynamics, debts do not rot with old age and are not consumed in the process of living. On the contrary, they grow at so much per cent per annum, by the well known mathematical laws of simple and compound interest.[14]

Economist Herman Daly explains the inevitable consequences that result when society pits the mathematical notion of compound interest against the physical reality of thermodynamics. He says that while debt can grow at compound interest forever, real physical wealth cannot continue to grow at the same speed 'because its physical dimension is subject to

145

the destructive force of entropy.'[15] Echoing Frederick Soddy's earlier analysis, Daly concludes that:

Since wealth cannot continually grow as fast as debt, the one to one relation between the two will at some point be broken – i.e. there must be some repudiation or cancellation of debt. The positive feedback of compound interest must be offset by counter acting forces of debt repudiation, such as inflation, bankruptcy, or confiscatory taxation, all of which breed violence.[16]

At every step in the entire production and exchange process, work is done; namely, energy is expended by both humans and machines. Part of that energy is absorbed into the product and part is wasted. This means that the more stages in the economic process, the more energy is lost. The same principle is at work in the production process as in the simple food chain described in Part 2. In highly industrial societies the stages of the economic process continue to proliferate, meaning more and more energy is dissipated all along the line; and the resultant disorders create even greater long-range problems for society.

Take, for example, your morning English muffin. As we will show in the next chapter, the very process of modern petrochemical agriculture used to grow the wheat is extremely energy inefficient. But once grown and harvested, the folly is compounded manyfold thanks to our national mania for processed food. Here are just some of the energy steps that go into making your English muffin. (1) The wheat is taken by a fossil-fuel-driven truck made of nonrenewable resources to (2) a large, centralized bakery housing numerous machines that very inefficiently refine, enrich, bake, and package English muffins. At the bakery, the wheat is (3) refined and often (4) bleached. These processes make for nice white bread, but rob the wheat of vital nutrients, so (5) the flour is then enriched with niacin, iron, thiamine, and riboflavin. Next, to insure that the English muffins will be able to withstand long truck journeys to stores where they will be kept on shelves for many days, or even weeks, preservative (6) calcium propionate is added, along with (7) dough conditioners such as calcium sulphate, monocalcium phosphate, ammonium sulphate, fungal enzyme, potassium bromate, and potassium iodate. Then the bread is (8) baked and placed in (9) a cardboard box which has been (10) printed in several colours to catch your eye on the shelf.

The box and muffins are placed within (11) a plastic bag (made of petrochemicals), which is then sealed with (12) a plastic tie (made of more petrochemicals). The packages of English muffins are then loaded into (13) a truck which hauls them to the (14) air-conditioned, fluorescent-lit, Muzak-filled grocery store. Finally, you (15) drive two tons of metal to the store and back and then (16) pop the muffins in the toaster. Eventually, you will throw away the cardboard and plastic packaging, which will then have to be disposed of as (17) solid waste. All of this for just 130 calories per serving of muffin.

Not only have tens of thousands of energy calories gone into the entire process, but medical evidence suggests that both the additives and the lack of fibre in refined breads may pose a serious hazard to your health. In the end, the energy that was added to the muffins at each step of the process was insignificant compared with the energy that was dissipated at each step of the process.

Of the total amount of energy used in the food system, less than 20 per cent actually goes into the growing of food. The other 80 per cent is consumed by the processing, packaging, distribution, and preparation of the foodstuff. Almost twice as much energy is used to process your English muffin (33 per cent) as was used to grow the grain it was made from (18 per cent).[17]

The food-processing industry is now the fourth-largest industrial energy user in the nation – after metals, chemicals, and oil. Some sources estimate that food processing currently consumes nearly 6 per cent of the country's energy budget. As far as the industry is concerned, apparently, the more the better. For example, between 1963 and 1971 the per capita food consumption in the United States increased by 2·3 per cent. But the tonnage of packaging grew by 33·3 per cent and the number of packages by 38·8 percent.[18]

Along with the growth in packaging has come a new

industry: an entire army of 'food technologists' now busy themselves making sure that our food supply is given just the right artificial colour, scent, flavour, texture, and so on. Nothing can be left to chance. As one food technologist puts it, 'It's hard to compete with God, but we're making headway.' Indeed they are. Some $500 million in synthetic chemicals is added to our food every year – 2,500 additives. In 1979, each American consumed an average of nine pounds of additives, nearly double the amount in 1970. Four million pounds of dyes wind up in the food supply annually, a full sixteen times the amount used in 1940. Today, we eat more synthetic and artificial foods than the real thing.[19]

Convenience and processed foods, which are promoted as ways to liberate the individual from the 'drudgery' of spending more time in the kitchen in food preparation, are in reality chaining humanity to the effects of ever greater entropy. The little time saved in the kitchen is more than outweighed by the amount of work time (human energy) given over to earning the money to pay for the increasing prices of the processed foods. Each step of the food processing takes energy, and as the energy flows through the food chain we witness a concentration of power in fewer and fewer food technology corporations, the decline in the healthfulness of the American diet, and the increased use of nonrenewable energy.

Food processing is representative of other major industries – such as petrochemicals, auto, truck and air transportation, and synthetic fibres – that grew up in the era of high energy flow. All appear to be generating greater value (more products, more 'convenience') while all the time they are actually squandering the precious energy resources of the planet. Again, the economic system fosters the illusion of creating a more ordered, more materially valuable world, because consideration is given primarily to value added or entropy decreases, but rarely ever to energy dissipation and entropy increases.

If the Entropy Law were fully acknowledged, society would have to face up to the notion that every time we use part of the stock of available matter and energy it means two things: first, that one way or another, the individual, the institutions, the community, or the society ends up paying more for the disorder created in making the product than the value derived from the use of the product; second, less energy is available to be used by other people and creatures sometime in the future. This reality flies in the face of the way we have viewed the world for the past several hundred years. The entire Enlightenment world view is inspired by the principles of Newtonian mechanics, Cartesian mathematics and Baconian scientific methodology. Capitalist and socialist systems attempt to organize the physical world on the basis of these basic conceptualizations. Central to all three ideas is the notion of absolute repeatability of observation (the scientific method) and the absolute reversibility of all processes (universal mathematics and mechanical processes). In the real world, however, nothing is observable in the same manner twice and no occurrence is reversible. The Entropy Law tells us that all physical reality unfolds in only one direction and that while there must be a $-T$ for every $+T$ in maths, there is no such reversibility in the physical sojourn of the world around us. It is indeed bewildering that we have been attempting to organize the world for these past few centuries on the basis of mechanics, mathematics and the scientific method, when the real world simply does not conform to the central assumptions of reversibility and absolute repeatability. The reality is that when we leave this world, we leave it less well endowed as a result of our presence. When we glorify high energy production, then, what we are really promoting is an ever greater consumption of the finite store of resources of the planet. Seen in this way, the gross national product is more accurately the gross national cost, since every time resources are consumed they become unavailable for future use.

Actually, the term *consumption* is a misnomer, for nothing is ever consumed. A thing is used, usually for a very short period of time, and then discarded. Any way you look at them, the statistics are mind-boggling. As a nation, we annually discard 11 million tons of iron and steel; 800,000 tons of aluminium; 400,000 tons of other metals; 13 million tons of glass; and 60 million tons of paper. Add to this 17 billion cans, 38 billion bottles and jars, 7·6 million discarded TV sets, and 7 million junked automobiles.[20] The figures are no less awesome on the personal level. In 1974 the average American used 10 tons of mineral resources, including 1,340 pounds of metal and 18,900 pounds of nonmetallic minerals. In a lifetime, each American uses on an average approximately 700 tons of mineral resources, including nearly 50 tons of metals. If we add fossil fuels and wood, the per capita usage more than doubles to 1,400 tons. And this amount excludes water and food needs.[21]

It has been said before that the world could not possibly support another America. Looking at these figures, it becomes apparent that even one America is more than the world can afford. Imagine if the entire world tried to produce and consume as Americans do. It has been estimated that a middle-class American lives a style of life that is equivalent to the work produced by 200 human slaves.[22] Buckminster Fuller refers to us as possessing 200 'energy slaves' that run on nonrenewable resources. Another way of looking at it is in terms of the number of calories needed to sustain life. An average human diet consists of 2,000 calories daily. Yet the amount of energy calories we individually consume every day – in our cars, our electricity, our processed foods, and so on – amounts to about 200,000 calories, or more than a hundred times the quantity we absolutely need.[23] In terms of energy consumption, though Americans number only 225 million people, our energy needs are equivalent to that of over 22 *billion* individuals!

It should also be understood that there is no way to allow for the needs of future generations in classical economic theory. When we meet as buyers and sellers in the marketplace we make decisions based on the relative abundance or scarcity of things as they affect us. No one speaks for future generations at the marketplace, and for this reason, everyone who comes after us starts off much poorer than we did in terms of nature's remaining endowment. Imagine what it would be like if all future generations for the next 100,000 years could somehow bid for the oil our generation is using up. Obviously, the price of that energy would be so expensive that it would be prohibitive if future generations were allowed to participate in today's resource allocation decisions.

The illusion of material progress is exemplified over and over again in every major economic and social activity simply because the second law is swept under the rug. Take, for example, the area of agriculture, transportation, urbanization, militarization, education, and health. In all six areas, we have convinced ourselves that we have made tremendous progress and that while there may be an occasional roadblock or retreat here and there, the progress is of a 'permanent' nature. On closer examination such claims turn out to be pure bunkum. The debunker turns out to be the second law.

In the following pages we will look at these six areas as typical case studies of the effect of the Entropy Law on economic and social activity. The pattern that each follows is duplicated over and over again in every area in contemporary society.

Agriculture

'Thank God for American agriculture!' Our agricultural system is the envy of the world. The bright yellow wheat fields stretching across mile after mile of Kansas flatland, the mechanized dairy farms scattered across the Wisconsin countryside, the lush fruit orchards planted along the entire rim of Southern California are all praised, studied, and copied by nations all over the world. Jonathan Swift once remarked that the man who can grow two ears of corn where only one grew before will deserve the better of mankind.[24] Can it be denied that American agriculture has succeeded beyond anyone's fondest expectations? Since 1940, crop output in the United States has grown at 2 per cent per year. In 1972, American agriculture recorded the highest yield figures in history.[25] Where else, said former Secretary of Agriculture Clifford Hardin, can one person raise 75,000 chickens in a modern, mechanized broiler-feeding system, or feed 5,000 cattle in an automated feedlot.[26]

Today, over 100 million people are starving to death all over the globe. Another 1½ billion people, nearly one-third of the human race, goes to bed malnourished each night.[27] With worldwide population expected to double in the next several decades, demand for increased food production will be greater than ever before in history. American agriculture is already producing 20 per cent of the world's wheat and feed grains and exporting over half of it to countries around the planet.[28] Certainly, looking at the statistics, one would be hard pressed to deny what everyone accepts as gospel: that American agricultural technology is extraordinarily efficient. Yet, the truth is that it's the most inefficient form of farming ever devised by humankind. One farmer with an ox and plough produces a more efficient yield per energy expended than the giant mechanized agrifarms of modern America. Hard to believe, but it's absolutely true.

A simple peasant farmer can usually produce about 10 calories of energy for each calorie expended. Now it's a fact that an Iowa farmer can produce up to 6,000 calories for every calorie of human labour expended, but his apparent efficiency turns out to be a grand illusion when all of the other energy expended in the process is calculated in. To produce 'just one can of corn containing 270 calories,' the farmer uses up 2,790 calories, much of which is made up of energy used to run the farm machinery and the energy contained in the synthetic fertilizers and pesticides applied to the crop. So for every calorie of energy produced, the American farmer is using up 10 calories of energy in the process.[29]

Today agriculture accounts for 12 per cent of all the energy used in the US economy.[30] Where farming traditionally relied on human and animal labour for cultivation, natural manures and crop rotation for fertilizing and maintaining the soil, and natural pest enemies for controlling crop damage, nowadays sophisticated machinery and petrochemicals have been substituted.

The more the energy flow-through has increased with the substitution of complex machinery and petrochemicals, the more centralized the agricultural industry has become. As the costs of maintaining the energy needs of US agriculture have escalated, the small family farmer has been literally driven off the farm and replaced by huge agribusiness corporations. Today, twenty-nine corporations own over 21 per cent of all of the cropland in America.[31] The next time you sit down to dinner, just think about this: your turkey probably came from Greyhound, your ham from ITT, your vegetables from Tenneco, your potatoes from Boeing, and your mixed nuts from Getty.[32] Agribusiness corporations now control 51 per cent of our fresh vegetables, 85 per cent of our citrus crop, 97 per cent of our broiling chickens, and 40 per cent of our eggs.[33] Only the large corporations can afford the mounting capital costs associated with a mechanized, energy-based agriculture. For example, it is estimated that the cost of farm machinery alone tripled between 1950 and 1971, from $12·1 billion to $33·8 billion in value.[34]

In that same time span, the use of inorganic nitrogen

fertilizer increased sevenfold – from 1 million tons in 1950 to 7 million tons by 1970.[35] The use of pesticides increased by even more.[36] These fertilizers and pesticides are all derived from fossil fuel energy sources. It would not be inaccurate to say that the food we eat today is grown from oil rather than soil. It is also true that it takes more and more oil to produce the same output of food each year. According to one authoritative study, it took five times more nitrogen fertilizer to 'sustain the same yield' of crop in 1968 as it did in 1949. In other words, five times more energy or work had to be expended to produce the same results.[37]

This is because, in agriculture as in everything else, every time energy is expended, some is absorbed into the product and some is dissipated. In order to increase yield, American farmers have continued to increase the amount of energy used. While some of that energy has helped increase output, more and more of it has been wasted altogether. The marginal entropy decrease, represented by slightly expanding yields, is dwarfed by the greater increase in dissipated energy in the overall environment. Much of the dissipated energy runs off and contributes to polluting our land, rivers, and lakes. Nitrate pollution from fertilizer runoff accounts for over half of our water pollution and two-thirds of our solid waste pollution.

Chemical pesticides are the other major energy input in modern agriculture. The use of pesticides increased from 200,000 pounds in 1950 to over 1·6 billion pounds in 1976.[38] A good part of the reason for this gigantic increase rests with the kind of farming technology we rely on. The United States has replaced diversified farming with monoculture farming in order to increase output. Monoculture crops are not well suited environmentally to attract the natural enemies of insect pests. In their absence, massive doses of chemical pesticides have to be used against the insects. The results, however, have been anything but successful. Studies have shown that even with the intervention of massive amounts of chemical pesticides, crop losses due to pest damage have remained at about one-third of total production for the past thirty years.[39] There is a good explanation for this. The pests have

154

developed genetic strains that are resistant to the chemicals employed. According to the annual report of the government's Council on Environmental Quality, there are now '305 species of insects, mites and ticks ... known to possess genetic strains resistant to one or more chemical pesticides.'[40] As the pest populations continue to develop more resistant genetic strains, even more chemicals of a more virulent nature have to be applied, which results in the development of even more virulent pests, the cycle becoming more costly and brutal at each stage.

The long-range effect of pesticide escalation on the ecology of the soil is 'frightening,' says agricultural expert Deryel Ferguson. Like others who have begun to study the problem, he warns that damage to soil by pesticides poses a threat of incalculable proportions. 'Every ounce of fertile soil contains millions of bacteria, fungi, algae, protozoa and small invertebrates such as worms and arthropods.'[41] Ferguson points out that all of these organisms play an essential role in maintaining 'soil fertility and soil structure.' Pesticides are destroying these organisms and their minute but complex ecological habitats, thus greatly hastening the entropy process of the soil. The end result is massive soil depletion and erosion. The use of chemicals, both pesticides and fertilizers, is partially to blame for the destruction of 4 billion tons of topsoil, which is washed into tributary streams each year.[42]

According to the Council for Agricultural Science, 'A third of all cropland is suffering soil losses too great to be restrained without a gradual, but ultimately disastrous, decline in productivity.'[43] The National Academy of Science now estimates that one-third of all valuable US farmland topsoil is already gone forever.[44] As the topsoil erodes, more chemical fertilizers have to be added just to make up for the deficit. In 1974, it would have taken $1·2 billion worth of chemical fertilizers to replace the natural nutrients lost through soil erosion.[45] Our farming technology, then, is caught in a vicious spiral of greater energy infusions in the form of fertilizers and pesticides and greater losses in the form of soil erosion and pest resistance.

As more and more energy is expended in American

agriculture, the entropy of the overall environment increases. The accumulating disorder in the form of pollution and soil erosion increases the overall cost for both society and the agricultural sector. The increased cost leads to the further enlargement and centralization of the economic institutions controlling agriculture. As these giant agribusinesses grow they require more and more energy just to maintain their operations, which means more energy has to be diverted away from the flow line. The increased costs of maintenance, of course, are passed on along the entire flow line. The final victim of the process is the consumer at the checkout line at the neighbourhood supermarket, who is forced to pay higher prices every week for the food – energy – needed to sustain life.

Every step of this high-energy agricultural process will continue to escalate as we move closer to the entropy watershed for nonrenewable fossil fuels. The frustration and anger experienced in petrol lines in the summer of 1979 was just a warmup exercise for what is going to take place in the grocery store lines in the years ahead.

Transportation

Our transportation system is supposedly the most advanced in the world. We spend more time on 'saving' travel time than on any other single economic activity. Transportation now accounts for 21 per cent of our gross national product.[46] Over 80 per cent of our transportation dollars are spent on autos and trucks.[47] Together, all of our major forms of transportation use up over 25 per cent of all of the energy consumed by the economy.[48] Even this figure grossly underestimates the amount of total energy consumed in transportation, because it does not include the cost of manufacturing and maintaining all of the transportation machinery. According to Dr William E. Mooz of the Rand Corporation, when these figures are added in, the American transportation industry eats up over 41 per cent of all of the energy used each year.[49]

Contrary to popular opinion, America's transportation system, like the agricultural system, has become less and less efficient over the years. That is, more and more energy inputs have been required to move the same amount of freight and passengers from one place to another.

Over the course of this century America's transportation system switched from primary reliance on the railroads to greater dependence on autos, trucks, and aeroplanes. Today, autos and trucks account for the most passenger and freight traffic respectively. Both are less efficient than other modes of transportation that have been either cut back or put into mothballs. It takes 8,100 BTUs of energy to transport one passenger one mile by automobile. In contrast, it takes only 3,800 BTUs of energy to transport the same person by mass transit.[50] Yet, over the past twenty-five years mass transit in this country has been greatly reduced. The figures for freight transport are even more telling. It takes only 670 BTUs of energy to ship one ton of freight one mile by train and over 2,800 BTUs to ship the same ton of freight by truck.[51] Even so train transport of freight was cut from 50 to 33 per cent between 1950 and 1970.[52]

All of our major modes of transportation run on nonrenewable fossil fuels. As the energy needs of America's transportation system have increased, the transportation industry has become more centralized in the hands of fewer companies. Where there used to be scores of domestic auto makers, today the industry is dominated by the Big Three auto firms – Ford, GM, and Chrysler. The same pattern occurred earlier with railroads, buses, and aeroplane traffic. Only these giant transportation firms can absorb the increased costs associated with the use of greater amounts of energy. Even they, however, are now feeling the crunch as the economy reels toward an entropy watershed. The auto industry, the undisputed leader of the American economy, is being forced to cut production and build smaller cars as the fuel crisis deepens. And as Henry Ford remarked, 'Mini cars make mini profits.'[53]

Smaller and fewer cars mean the entire economy suffers. Autos consume '20% of all the steel, 12% of the aluminium, 10% of the copper, 51% of the lead, 95% of the nickel, 35% of the zinc, and 60% of the rubber used in the US.'[54] Way back in 1932, one auto zealot summed up the great possibilities for the entire economy in expanding auto production:

Think of the results to the industrial world of putting on the market a product that doubles the malleable iron consumption, triples the plate glass consumption, and quadruples the use of rubber! . . . As a consumer of raw material, the automobile has no equal in the history of the modern world.[55]

In 1974 Americans spent $137 billion on private auto travel.[56] Every twenty-four hours 10,000 new drivers and 10,000 new cars are added to the road.[57] The American consumer spends one out of every four dollars on the automobile.[58] He pays for the car, the insurance, the petrol, the maintenance, the parking charges, the highway tolls, the traffic tickets, the state and federal taxes, and by the time he's done he has spent more money than he spends on food.

Today, one out of every six jobs is directly or indirectly related to the automobile.[59] The automobile is a central

158

feature of our fossil fuel culture. The increased expense, then, in buying, running, and maintaining an automobile is a good measuring stick of the increased costs incurred all along the energy flow line as we run up against the end of the age of fossil fuels. The massive disorders caused by the automobile are also a good example of what happens when an economic system fails to take into account the effects of the Entropy Law until it's too late. Whatever we have received in benefits from the automobile over the past fifty years must now be judged in light of the even greater penalties we are now being forced to pay as the second law relentlessly drives its point home. The total bill is more than any of us can afford, as a brief survey of some of the costs suggests.

The first cost to consider is time itself. The automobile was supposed to reduce the amount of time it takes to get from one location to another. In truth it has done the opposite. With the widespread use of autos, Americans began to move farther away from their place of work. Forty years ago, most people lived within walking distance of their place of employment. Today people are spread out in suburbs, sometimes twenty or thirty miles from their job. While the automobile is a faster mode of transportation than walking, its speed becomes relatively meaningless when peak rush-hour traffic crawls at a pace of five to six miles an hour, as it now does entering and leaving many of America's major cities. It now takes most commuters anywhere from thirty minutes to an hour and a half each way to get to and from work – about the same time it took people forty years ago when homes were located nearer to the jobs and people could walk or take a trolley car. With the fuel crisis, even more time is wasted in the automobile. In the summer of 1979, automobile owners in many areas of the country were forced to spend between one and four additional hours per week just waiting in petrol lines to get a fill-up.

Former Secretary of Transportation Alan Boyd once remarked:

If someone were to tell you he had seen strings of noxious gases drifting among the buildings of a city, black smoke blotting out the sun, great holes in the major streets, filled with men in hard hats, planes circling overhead, unable to land, and thousands of people choking the streets, pushing and shoving in a desperate effort to get out of the city . . . you would be hard pressed to know whether he was talking about a city at war or a city at rush hour.[60]

In fact, the death and destruction wrought by the automobile is more gruesome than anything our country has ever faced in wartime. Automobile accidents kill 55,000 Americans each year and maim 5 million others.[61] The National Safety Council estimates that more Americans have been killed by automobiles than were killed in *all* of the wars this country has fought in the past 200 years. Imagine, over 1 million people have been killed by automobiles in just the past thirty years![62]

In dollar terms, the loss of health and property incurred by traffic accidents is ten times the total from all other crimes of violence combined. In 1969 the total losses from traffic accidents rang up at about $13 billion. By 1975, the societal costs of losses involving auto accidents reached $37 billion.[63]

Even these losses are only part of the picture. With the auto age came highways, and thousands and thousands of miles of concrete, asphalt, and cement. The environmental damage resulting from the lethal combination of highways and motor cars is phenomenal. The first portland cement concrete was used to pave a small stretch of public road from Detroit to the Wayne County State Fairgrounds in 1909.[64] From these humble beginnings America set out on a public works project which has become the most costly ever undertaken in the history of the world. Between 1956 and 1970 alone, this country spent $196 billion in local, state, and federal monies on highway construction. State governments estimate that between 1973 and 1985 they will need an additional $294 billion for highway-related work.[65]

The interstate highway system takes up 42,500 miles.[66] 'The highway system,' says transportation expert George W. Brown, 'devours land resources and atmosphere at a rate that is impossible to sustain.'[67] According to the National Highway Users Conference, for every million dollars spent on the interstate highway system, 16,800 barrels of cement are consumed, along with 694 tons of bituminous materials; 485

160

tons of concrete and clay pipe; 76,000 tons of sand, gravel, crushed stone, and slag; 24,000 pounds of explosives; 121,000 gallons of petroleum; 99,000 board feet of lumber; and 600 tons of steel.[68]

There are 3,600,000 square miles of land in the United States and over 3,600,000 miles of road: that's one mile of road for every square mile of land.[69] Roads are proliferating so fast that they now take up 30 per cent of the land use in fifty-three central cities.[70] Approximately two-thirds of the downtown land area of Los Angeles is now devoted exclusively to either the parking or movement of automobiles. In Chicago, Detroit, and Minneapolis nearly half the city land is 'devoted to the movement and the storage of automobiles.'[71] While there is no way to calculate the total damage done to buildings and other city structures by the constant friction, weight, movement, and general wear and tear produced by inner-city auto traffic, city planners have begun to introduce the phrase 'auto erosion' into their vocabulary when studying the cost factors of city traffic.

The small entropy decrease represented by every mile of highway and every shiny new car that travels along it is bought at the expense of a tremendous increase in entropy in the overall environment. Anyone who's been unfortunate enough to live right smack in the path of a highway construction route has felt the effects of the second law personally. According to D. R. Neuzil of the Institute of Transportation and Traffic Engineering of the University of California, close to 100,000 persons per year are uprooted and displaced by new highway construction.[72] The destruction of neighbourhoods has taken its toll in countless unforeseen ways, as sociological studies now confirm. The breaking up of long-established human living habitats has had an effect every bit as destructive as the breaking up of biological habitats. The resulting disorders have often been reflected in an increase in crime, unemployment, and mental illness, as familiar patterns of life have been suddenly and traumatically altered. Think of what it must do to the human psyche to see one's entire neighbourhood of several square blocks suddenly razed to the ground. The sense of utter loss and confusion, say the psychologists, is

161

often similar to that experienced after the ravages of bombing raids during wartime.

Finally, there is pollution to consider. Every time one of America's 150 million automobiles (or trucks or buses) travels along the highway, it is expending energy, much of which is dissipated as carbon monoxide, nitrogen oxides, and hydro-carbons. Today, 60 per cent of the total air pollution in most US cities is caused by auto exhaust.[73] In 1971, damage to buildings and property due to air pollution was estimated at $10 billion.[74] It is now acknowledged that the dramatic rise in deaths caused by heart disease and cancer is also partially traceable to air pollution caused by the exhaust fumes of cars, trucks, and buses (more about this in the section on health).

Every day '250,000 tons of carbon monoxide, 25,000 tons of hydrocarbon and 8,000 tons of oxides of nitrogen' are spewed out from auto exhausts. In 1970 the auto pollution totalled 111 million tons of sulphur oxides, 19·5 million tons of hydrocarbons, and 11·7 million tons of nitrogen oxide.[75]

What's really scary is the recognition that the entropy process often affects activities so far removed from the original energy expenditure that no relationship is even suspected. For example, a man driving an automobile down the highway would, no doubt, be shocked to learn that every time he puts his foot down on the accelerator he is potentially contributing to the brain damage of a five- or six-year-old schoolchild miles away. Studies over the past several years have demonstrated that children with learning disabilities, who show signs of 'drowsiness, irritability, abdominal pains and vomiting . . . and in severe cases – paralysis, convulsions and coma,' also have higher concentrations of lead in their blood.[76] Most of the lead poisoning comes from auto ex-hausts. According to the findings of one detailed study undertaken by the Children's Hospital at Harvard Medical School:

Teachers rated the classroom behaviour of 2,146 of the children without knowing the lead levels of any of them. When these behaviour ratings were later correlated with the findings from studies of the shed teeth (baby teeth), the scientists found a direct and regular relationship. The higher the level of lead contamination, the

162

more likely were behavioural problems such as lack of persistence and organization, failure to follow simple directions and tendencies toward lack of attention, impulsiveness and excessive activity.[77]

Today's high-energy transportation system is more responsible for tearing up our society and depleting our energy base than any other single force. Its destructiveness cannot be sustained much longer without doing irreparable damage to our society and threatening our very ability to survive as a nation.

Urbanization

'It's no news that America's major cities, on which much of the nation's growth depended for the last 200 years, are in decline. The challenge is whether this decline can be halted, or whether *all* big cities are to falter and eventually become ghosts of their once-thriving selves.'[78] Foreboding words, to be sure; the kind you might expect from a radical sociologist or a 'small is beautiful' advocate. But the message here is enhanced by its source, *US News and World Report*, the most business oriented of the nation's newsweeklies.

Since World War II, and the advent of fossil-fuel-based agriculture, America has been an urbanized country. Today, 80 per cent of the population lives in urban areas. More than half of our people inhabit just 1 per cent of the land.[79] Between 30 and 40 million of us live on just 10,000 square miles sandwiched in between southern New Hampshire and northern Virginia.[80] For a long time the city symbolized greater opportunity, more jobs, higher culture. No more.

Today, great numbers of Americans are becoming increasingly disenchanted with large cities. Recent polls have shown that most people actually want to live in smaller communities. 32 per cent would prefer to live in small towns or cities; 25 per cent in suburban communities; 26 per cent in rural areas; and only 17 per cent in big cities.[81] These sentiments are being translated into action. Between 1970 and 1976 the top seventeen metropolitan areas in the country actually suffered a net loss of nearly 2 million people.[82] When asked why they are leaving our great metropolises, the out-migrants are likely to respond in a variety of ways, citing crime, taxes, food and housing costs, crippling strikes of municipal workers, loss of jobs, pollution. All of these people are reacting to different facets of the same phenomenon: because of the massive energy inputs required to sustain contemporary city life, the entropy of the urban environment is rising dramatically, to the point where the continued existence of urbanization is being called into question.

The city as we understand it – millions of people packed together into giant megalopolises sprawling over hundreds of square miles – is a relatively new social institution, dating back to the birth of the era of fossil fuels. Before the rise of the modern urban centre, people had lived in 'cities' for thousands of years. But these were hardly cities at all by modern standards. Ancient Athens, for instance, had just 50,000 citizens; Babylon, a little more than 100,000. Centuries later, during the Renaissance, the size of the urban areas had changed very little. Leonardo da Vinci's Florence claimed 50,000 residents, and when Michelangelo painted the ceiling of the Sistine Chapel, Rome's entire population hovered around 55,000. As late as the sixteenth century, the majority of European cities housed fewer than 20,000 inhabitants. At the time of the American Revolution, the two largest cities in the colonies – Boston and Philadelphia – had not yet reached 50,000, and New York City ranked a distant third in size.[83]

With the spread of the Industrial Revolution in the early nineteenth century, all of this began to change overnight. London became the first city with a population of 1 million in 1820. By 1900 there were 11 cities with populations exceeding 1 million; by 1950, 75 cities; by 1976, 191 urban areas composed of 1 million or more people. At present worldwide growth rates there will be 273 cities with populations over a million by 1985, the majority of these in Third World countries.[84]

As a percentage of the world's population, urban dwellers are moving toward a majority. Of the estimated 1 billion people alive in 1800, perhaps 25 million of them – or just 2·5 per cent – lived in cities. By 1900, 15 per cent of the world's population was located in urban areas. By 1960, one-third of the population. At the current growth rate, by the year 2000 more people will live in cities of 100,000 or more than lived in the entire world in 1960.[85]

This incredible explosion of urban life has come as a direct result of the global shift in the energy environment during the past two centuries. A city survives by virtue of its ability to

165

gather available energy from the surrounding environment and store and use it for urban existence. Cities first originated thousands of years ago with the discovery of hard-grain cereal cultivation. Hard grains, unlike fruits and vegetables, lend themselves to long-term storage. Surveying the cities of the pre-fossil-fuel era, we can designate urban areas on the basis of their energy base as 'rye cities, rice cities, wheat cities and maize cities.'[86]

While hard grains provided the energy foundation for urban life prior to 1800, they also set severe restrictions on both the size of the city's population and the physical size of the city itself. Traditional agriculture could not yield a large enough surplus to support a massive urban population of non-food producers. Because the city was directly reliant on the surrounding countryside to provide it with energy (food), urban areas could not sprawl over the landscape as they do today for fear of destroying their local food base. The military walls surrounding ancient and medieval cities provided more than protection from invasion; they also ensured that the city would not grow beyond the limits of the carrying capacity of the energy environment. The great city of Babylon, for instance, encompassed an area of just 3·2 square miles; the medieval walls of London enclosed an area less than 1/150 the city's present size. Nor could the traditional city rely upon food supplies brought in from great distances. Until the fossil fuel era, most transport was conducted through either animal or human labour, and thus the society's energy base set an absolute limit on the speed and distance over which food could be hauled.

The well-known exception to these historical limitations was the ancient city of Rome. At its peak, it grew to a population of nearly one million people. The Roman city could only be sustained, however, by attempting to colonize everything in its path. Without its vast pool of slaves, intensive farming techniques, massive aqueduct-building projects, and, most importantly, the empire's armies, Rome could not possibly have supported its population. In a sense, the entire known world had to be pillaged to overcome the natural limitation imposed by a solar-agricultural energy base.

Murray Bookchin puts it well when he writes, 'The Fall of Rome can be explained by the rise of Rome. The Latin city was carried to imperial heights not by the resources of its rural environs, but by spoils acquired from the systematic looting of the Near East, Egypt and North Africa. The very process involved in maintaining the Roman cosmopolis destroyed the cosmopolis.'[87]

Once embarked on the course of urban expansion, Rome was in a losing race. The larger the city became, the more energy inputs were required. The more energy flowing into the city, the greater the resulting disorder. The greater the disorder, the larger became the institutional infrastructure to deal with the various types of chaos. The process simply could not be sustained indefinitely. The energy supply lines maintained by the army became stretched so thin that the military absorbed more energy than it returned to the city. The agricultural system began to experience diminishing returns because of the intensive abuse of the soil. Slaves became too expensive to feed and house. The city's bureaucracy grew so big and costly that it could not be supported. Eventually, the overbloated city collapsed from within and without, returning after its military conquest to ecological equilibrium with its energy environment. After its fall, Rome claimed just 30,000 inhabitants.

Rome serves as a case study of what can happen when an urban area vainly seeks to ignore the growth limitations imposed on it by its surrounding resource base. Seeking out far-flung energy resources can serve to delay the collapse, but eventually the day of reckoning must come. Such is the case in our own time. Modern urban areas are supported through a kind of colonization of the world that is quite similar to that which sustained Rome. And like Rome, modern cities, because they have far outstripped the productive capacity of their local energy environments, are extremely vulnerable to collapse once the limits of their national and international resource base are reached.

Nowhere is this more evident than in the modern city's needs for food. A typical urban area of 1 million people requires a daily input of 4 million pounds of food.[88] To get these 2,000 tons of nourishment, the city is completely reliant upon our fossil-fuel-based agricultural system. Without the high yields of petroleum-chemical farming, and a national transportation system that moves wheat and oranges and beef thousands of miles to scattered urban areas, major cities would quickly become scenes of mass starvation. But as we have just seen, the declining availability and escalating cost of fossil fuels – the backbone of American farming and transportation – threaten the survival of the very agricultural system upon which the city is dependent.

Where will the food for New York, Chicago, and Los Angeles come from? Not from the surrounding countryside. Due to urban and suburban sprawl, tens of millions of acres of potential food-yielding land have been converted to concrete, plastic, and steel.[89] And not from within the city itself. In the historical city, a fair amount of land within the city walls would be set aside for small-scale agriculture. But as cities have grown bigger, more and more potential food-growing land has been converted to other uses. Fully half of Dallas, for example, is covered by roads, parking places, and buildings. In the entire 319 square miles of New York City, just 30,000 acres of unused land remained by the mid-1950s.[90]

Major urban areas are precariously reliant on other types of far-flung resources as well. A city of a million requires a daily input of 9,500 tons of fuel and 625,000 tons of fresh water.[91] Construction and maintenance of America's buildings (most of which are in large urban areas) require 57 per cent of all the electricity produced in the country. Lighting them alone takes about one quarter of the nation's electricity.[92] Mammoth buildings like the World Trade Center draw 80,000 kilowatts of electricity, enough to service the entire city of Schenectady, New York. The Sears Building in Chicago uses more electricity than the people of Rockford, Illinois, a city of 147,000 inhabitants. Massive inputs of natural resources are also required. The Sears Building, for instance, contains eighty miles of elevator cables and enough concrete to cover an area equivalent to seventy-eight football fields.[93] Resources are also necessary for upkeep. In cities

168

around the country, steel is deteriorating so rapidly that the current annual replacement cost is estimated at $20 billion.[94]

Without the massive inputs of energy in its various forms, the city decays, jobs are lost, and urban life becomes intolerable. This process is already far advanced in some of the nation's oldest cities. According to a study done by the Urban Institute on the condition of America's urban infrastructure, basic facilities like sewers, streets, bridges, transit systems, and waterworks in the nation's major cities are finally wearing out and will require massive expenditures of funds to replace or repair in the coming decade. The figures are truly staggering. To avoid the collapse of its physical plant, New York City will require the expenditure of $12 billion for replacement, repair, and maintenance operations over the next ten years. Even a smaller city like Cleveland will have to spend over $700 million if it is to maintain its physical infrastructure in the years just ahead.[95]

Big cities require big inputs of energy to remain viable. As the energy flows into the urban area, however, it undermines the vitality of the city by generating various disorders. For example, a high energy flow into a city causes significant ecological changes. A large city's annual temperature averages three or four degrees hotter than surrounding areas. This is due to the emissions from power plants, automobiles, air conditioners, and the changes in solar reflection caused by highways and buildings. There are ten times more air pollutants in the city as in rural areas. Other meteorological phenomena created by an urban area's energy requirements include: 100 per cent more winter fog and 30 per cent more summer fog than in surrounding rural areas; 5 to 10 per cent more rain and snow in the city; 5 to 15 per cent less sunshine; and 20 to 30 per cent less wind.[96]

The high levels of energy consumption and the resulting waste in cities seriously affect the health of urban residents. City dwellers have inordinately high cancer rates, along with more bronchitis, ulcers, and heart disease. The residents of large cities also exhibit far more antisocial behaviour, hostility, and selfishness than those who live in lower-concentration energy environments. The suicide rate is higher in big cities;

the percentage of admissions to mental hospitals is greater; schizophrenia, neurosis, and personality disorders are all considerably higher in urban environments. The crime figures alone are astonishing: there are 5·7 murders per 100,000 people in cities of 25,000–50,000 residents, but 29·2 murders per 100,000 in cities of over a million. A city of 100,000 averages 300 violent crimes annually; a city of over a million, 11,880.[97]

The density of high-energy urban life can affect human relationships and interaction in a more subtle way. For example, it has been estimated that a person can 'meet' 220,000 people within a ten-minute radius of any place in midtown Manhattan. Obviously, it is impossible to pay attention to each person, so urban dwellers establish a kind of screening process, giving less time and consideration to each input. Urbanites typically disregard 'low-priority' inputs such as panhandlers and drunks. Dozens may witness a crime and not report it or aid the victim. A simple walk down a street becomes a process of assuming an unfriendly face to ward off people who are undesirable 'inputs.' To preserve psychic energy, people in large cities become friendly with far fewer people than do those who live in sparsely populated rural areas. Neighbours are often totally anonymous. We become like sailors in a lifeboat: everywhere we are surrounded by water, but not a drop to drink.

Highly urbanized life tends to destroy effective political participation. In a small town, anyone might drop in to see the mayor to discuss a local issue. But in a major city, the individual's opinion and participation become nearly meaningless. A member of the New York City Council represents an average of 239,000 people. If he spent eight hours a day, every day of the year, doing nothing but talking for fifteen minutes with each of his constituents, a council member could only talk to 10,000 of them over the course of a year.

Kirkpatrick Sale, in a study analysing the quality of life in large cities (of over a million) versus the small city (under 100,000), argues that in any area we care to look, major urban centres are inferior to small, decentralized communities. Not only are big cities vulnerable to massive unemployment during times of economic crises, but, on a daily basis, 'there are higher transportation costs because of congestion, higher employee sickness and death rates because of air and water pollution, higher maintenance and cleaning costs because of air pollution, higher energy costs because of the "heat island" effect over cities in the summer (and the inaccessibility of dense buildings to sunlight in the winter), higher security costs and higher loss rates because of crime, higher costs in training new workers because of bad schools.'[98]

Urban expansion means higher energy flows and mounting disorders. As the various disorders build up, the city bureaucracy grows in an attempt to impose some order on the developing chaos. Still, every major city has discovered that there is just no way to adequately provide the necessary services – power, sewerage, schools, highways, police, public housing, and so on – that are required. One study indicates that the service demands upon a large city double every year. In New York City the number of municipal workers increased by 300 per cent in the last decade, while the city's population actually declined.[99]

Obviously, the energy put into the city must also come out in the form of waste. The garbage problem in any major urban area is truly monumental. In metropolitan Washington, DC, 4,000 tons of garbage are collected and compacted every twenty-four hours. If this daily accumulation of waste were dumped on the Mall in downtown DC, it would stack up nearly half as high as the top of the Washington Monument. Where does all of this garbage go? In DC, the urban area has five major landfills where waste is discarded. All five of these fills are beginning to spill over. Of course, more dumping sites could be built, but because the metropolitan area is so densely populated, any new landfill site would inevitably have to be placed near where thousands of people live. While everyone wants his garbage picked up and hauled away, no one wants a garbage dump built near his home. Faced with this problem, city authorities have two choices. Either the garbage will have to be burned, which will mean dirtier air

171

and more pollution, or it will have to be packed into railroad boxcars and shipped to less populous areas of the country, a process which will use a considerable amount of energy and cause higher city income taxes.[100]

Maintaining a high energy flow-through and absorbing the increasing disorders that build up along the city's flow line require money. The Urban Institute has shown that a resident of a city of 1 million people will typically pay three times more in taxes than a resident of a small city of 50,000 inhabitants.[101] The bulk of this money will go into education, police, and health services. Yet by all statistical measurements, urban residents suffer more crime and have more inferior schools and worse health than those who live in small cities or rural areas. The entropy of the urban environment continues to build as a result of increased energy inputs, and the city's problems become unsolvable in conventional terms. Notes economist Leopold Kohr, 'Social problems have the unfortunate tendency to grow at a geometric ratio with the growth of the organism of which they are a part, while the ability of man to cope with them, if it can be extended at all, grows only at an arithmetic ratio.'[102]

Eventually, the city begins to run out of available resources and reaches the point where it is spending itself into bankruptcy. According to the Center on Environmental Quality, 'expenditures in most distressed cities are growing much faster than increases in the value of real property, the chief tax base in most municipalities.' In a city of 1 million or over, the average revenue per capita equals $426·90 in taxes, but the average local debt incurred by the city to pay for services equals $1,052 per resident.[103]

Even as the city attempts to preserve itself, it actually fosters its own economic decline. Rising taxes induce wealthy and middle-class residents, along with corporations, to leave the city. As the wealthy and middle class leave the city, less tax revenue is available to the bureaucracy and fewer jobs remain. Unemployment rises, crime goes up, and the city is forced to spend even more to keep down the disorder. The vicious cycle continues on and on.

The near fiscal collapse of New York and Cleveland is a

sign of what lies ahead for our overgrown and outworn cities in the next two decades. The sober truth is that we can no longer afford to maintain these incredibly entropic urban environments.

The Military

The world has never seen a military machine the equal of our own. Of every dollar spent by the federal government, 43 cents goes to pay for wars – past, present, and future.[104] A recent military budget (fiscal 1980) called for $138 billion for the nation's defence for just one year, a $10 billion increase over the previous year. America's armed forces now include at least 25,000 nuclear weapons, 2 million soldiers, 500 massive naval ships, 10,000 aircraft, and 400 domestic military bases.[105] Twenty thousand military contractors work to produce tens of thousands of different weapons systems. Counting workers employed directly under contract to the Defense Department, more than 5 million of our citizens owe their livelihood to the Pentagon.[106]

As far as most Americans are concerned, this is just as it should be. From the day the Japanese bombs fell on Pearl Harbor, most of us have felt that the more military spending we have, the greater our national security. Remembering the darkest days of the Depression, many people recall how Franklin Roosevelt brought us out of it through massive military investment. Thus, defence spending has appeared for decades to be good for the economy. Yet, by any measure we care to use, it is becoming increasingly clear that the more resources and energy are devoted to the military, the less real wealth and security exists. As noted defence analyst Seymour Melman has written, 'Far from being dependent on arms production for our prosperity . . . we paralyse the country as a whole by diverting the lion's share of our resources into the military sphere.'[107]

Today, the US military is the largest single institutional consumer of energy in the nation. Over 80 per cent of the federal energy budget goes to the Defense Department.[108] When industrial defence contractors are included in, the military uses 6 per cent of the nation's total energy require-

174

ment.[109] Along with the energy has gone material and people. Since World War II, the defence establishment has been the largest single institutional user of capital and technology in the country. To sustain this infrastructure, fully one-half of the scientists and engineers of the past generation have worked either directly, or under military contract, for the Department of Defense.

The energy drained from society by the military causes tremendous social dislocation. Nowhere is this more readily visible than in the monthly unemployment figures. The common myth is that defence spending creates jobs. In fact, a study conducted by the Michigan Public Interest Research Group concludes that for every $1 billion added to the military budget, the nation as a whole loses 11,600 jobs. The study also found that in each of twenty-six states, which comprise 60 per cent of the nation's population, as military spending in the state rose, unemployment rose as well.[110] The International Association of Machinists, in its own recently concluded survey, showed that 'a Pentagon budget of $124 billion costs the Machinists over 118,000 civilian jobs. When the 88,000 jobs generated by this level of military spending are subtracted, the net job loss to the union is 30,000 jobs a year.' Another report, authored by Marion Anderson and released in 1978 by Senator Edward Kennedy, indicates that a military budget of $124 billion (1979's total) costs the jobs of 1,440,000 Americans.[111]

While it may appear paradoxical that investing money in military production actually produces unemployment, the issue is quickly resolved when we look at the nature of the jobs created. The types of jobs generated through military spending are necessarily highly capital and energy intensive. Human labour represents a very small ingredient in the overall mix of factors involved in weapons production. For example, the federal government has awarded Lockheed Corporation a twenty-year contract that provides $1 billion a year for the company to work on the most lethal and expensive weapon ever devised, the Trident submarine. The company employs 16,000 people on the project. This same $1 billion, however, could be used to create 20,000 jobs in more labour intensive, and less energy consumptive employment, such as the construction of solar collectors.[112]

Military spending is also a leading contributor to inflation. As the *New York Times* notes: 'Virtually all economists agree ... that military spending tends to be inflationary. This is because it puts money into the hands of workers without expanding the supply of goods they can buy – the consumer market for missiles and the like being somewhat limited – thereby driving up the prices of goods like autos and refrigerators and machine tools.'[113] Military production also causes inflation in a more important sense. The first law of thermodynamics tells us that the quantity of energy and matter is fixed. Because the military sector consumes 6 per cent of the nation's total energy use, along with massive quantities of nonrenewable mineral resources, the entropy increase represented by military hardware (the amount of energy no longer available to do work) causes tighter resource supplies, which in turn fuels inflation.

The rejoinder to all of this, of course, is that while energy-intensive military spending may indeed cause social disorder in the form of unemployment, inflation, and resource depletion, it at least provides us with a system of national security unparalleled in history. If security is to be measured in numbers only, then we would surely be the most secure nation on earth. If the lethal capacity of the nuclear arsenals of the world were broken down into tons of TNT, each man, woman, and child alive at this moment – nearly 4.5 billion of us – would be represented by four tons of explosive power. Some of our individual hydrogen bombs represent so much megatonnage that they each total more tons of dynamite power than all the bombs dropped by all sides during all of World War II. Using our atomic storehouse, we could blow up every major Soviet city fifty times over. And we daily produce two more nuclear bombs to add to our stockpile.[114]

Every dollar spent on national defence only generates greater global tension. Each time the United States develops a new weapon system, the Soviets feel threatened and therefore generate another as counterbalance. This, in turn, causes us to respond, ad nauseam. Today, we spend three times as much in real dollars as we did on defence in 1948,[115] but who could claim that we are three times more secure when, within twenty minutes of the commencement of an all-out nuclear war, 160,000,000 Americans would be dead?

Ever more sophisticated weapons systems mean greater energy concentration and flow-through. If the history of warfare teaches us one simple truth, it is this: the more concentrated the energy flow-through, the more deadly and depersonalized warfare becomes. At this moment, the Soviet Union and the United States are annually spending a combined total of $20 billion on the development of new weapons of war. The United States alone is experimenting with some 20,000 future weapons concepts.[116]

As American weapons systems become more energy intensive, more complex, and more expensive they suffer increasingly serious operational problems. Cost overruns become enormous and commonplace: the Trident submarine is currently $400 million over estimated costs. Some systems just plain don't work. The Department of Energy has admitted that 75 per cent of all Polaris A1 missiles (a popular military item in the mid-sixties) would not have functioned had they been fired. Unexplained crashes of the most technologically advanced planes have become a regular occurrence.[117]

Finally, as the military seeks ways to embody more and more energy in destructive devices, weapons systems grow to such complexity that they approach the ludicrous. A current military proposal, much favoured on Capitol Hill, is the MX. The idea here is that 200 missiles will be hidden from 'enemy' view by shuttling each warhead underground between twenty-five shelters. The theory is that the Soviet Union will never know where each individual missile is located, and so will have to launch 5,000 warheads to be certain to destroy them all. To construct the underground railroads and storage shelters necessary for this scheme, the Air Force will have to acquire an estimated 3½ million acres of land west of the Mississippi. This is an area roughly four times the size of Connecticut – just to house 200 missiles![118]

According to the governor of Kansas, a state in which the Pentagon hopes to build an MX tunnel, construction will disrupt 186,000 acres of prime western Kansas farm and range land; exclude the area's 40,000 inhabitants from 6,500 square miles of the state; suspend the existing uses of land in the area from farming, recreation, grazing, and human habitation for twenty to thirty years. On top of it all, construction crews and their families will bring in an additional 81,000 people to western Kansas, causing public service costs to rise by $37·5 million. And, asks the governor, how will the missiles be transported to their underground havens? Each weighs approximately 1 million pounds and is 150 feet by 22 feet in size. The entire cost of this Rube Goldberg scheme: $30–40 billion.[119]

The MX is just a small part of the various military systems being developed. Like something out of science fiction, the next generation of wars will be fought with missiles that read maps and guide themselves, and with killer satellites and particle-beam death rays hurled from outer space. By 1985, the United States hopes to have a fully operational high-energy laser which can melt tanks or focus its beam to disable a satellite orbiting a thousand miles above the earth.

Yet, for all of our destructive capability, perhaps never has a nation's military supremacy been so precariously perched. While soldiers of the past moved on their stomachs (food being their primary energy source), today's military runs on oil. And oil is a diminishing resource. The total Department of Defense energy costs in 1978 exceeded $4 billion, more than twice the costs in 1973. Rising energy costs have 'already caused the Defense Department to incur a lower margin of readiness than we might otherwise prefer,' according to the Pentagon's Ruth Davis in a testimony before Congress. Though DOD has managed to realize a 30 per cent reduction in energy usage since 1973, the savings were accomplished by reducing men and operations. Says Davis: 'Any further energy savings through reductions of this nature will have a serious impact upon our ability to maintain acceptable levels of force readiness.' The oil embargo of 1973 and the revolution in Iran in 1979 'provide renewed awareness of our increasing susceptibility to the potential for political, economic or military pressure – pressure applied by those who

178

either have the ability to control directly or who can indirectly influence the flow of oil to the US and to its allies.' As things now stand, the Pentagon believes that 'significant shortages of petroleum fuels for US needs will probably occur in the late 1980 or early 1990 time frame.'

Even though the energy supply needed for our military operations is dependent on foreign sources, our war machine has become so specialized that there is very little that can be done to provide alternative sources of energy. Ninety per cent of the DOD's daily use of petroleum is for 'mobility fuels,' used in aircraft, missile systems, and ships. 'The DOD continues to design and build weapons systems under the implicit assumption that they can be fuelled with petroleum-like products,' says Davis. Until 'well into the 21st century,' she adds, the military will have to rely on liquid hydrocarbon fuels.

To cope with its declining fuel supply, the military must furiously scramble to develop new sources. In the old days, the troops might simply be sent into the Middle East oil fields; in fact, the Pentagon recently surfaced just such a proposal to test public reaction. But any such incursion might well lead to nuclear war, so the military has been forced to look closer to home. Already, the President has guaranteed that the armed forces will continue to receive 100 per cent of their fuel requirements. This will clearly result in the further draining of energy away from other social and economic needs. The Pentagon has also become a prime motivator in launching the government and industry drive toward synfuels. If the synfuel industry is developed, the military hopes to consume 50 per cent of all synthetic fuels produced domestically. To keep its energy flowing, the Pentagon has even recommended the establishment of a Defense Mobility Fuels Action Plan which would give the military unprecedented control over the nation's energy policies. Given the declining availability of petroleum, any attempt by the Pentagon to maintain its own high energy flow is certain to cause greater disruptions in other sectors of society.[120]

As our weapons systems become more complex, and our military presence in the world expands, more and more energy must be used up just to maintain the growing military bureaucracy. According to Earl Ravenal, former DOD analyst, 'Less than 30% of our entire defence budget goes toward the direct defence of our country and its essential interests.'[121] The rest is consumed by our national attempt to maintain a worldwide military presence. In the current military budget, $35 billion was spent on the building of new weapons; the other $100 billion plus was used essentially for personnel and maintenance costs.[122]

Increased military spending represents a tragedy of monumental proportions. As the military fights to maintain itself and its own energy flow, it continues to take energy away from society's flow line, thus exacerbating other energy-based problems such as hunger and poverty. All of the nations of the world combined are currently expending $400 billion annually on weapons; nearly $1 million a minute.[123] Wars and preparation for wars consume roughly 10 per cent of the world's total production of all goods and services.[124] This is the equivalent of the entire GNP of over half the world's population. With 800 million people barely surviving on $200 or less annually, and with 20 million dying of hunger each year, high-energy military spending becomes an obscenity. If just 2 per cent of the world military budget for just one year were diverted, it could provide every rural Third World family with a stove.[125] Here in the United States, the cost of a single aircraft carrier – $1.6 billion – is almost twice the entire budget for occupational health and safety programmes; the unit cost of an A-7E Corsair attack plane equals two times the 1977 EPA budget for safe-drinking-water programmes.[126]

In the end, warfare, and its preparation, are the most highly entropic form of human activity. After all, there are only two things you can do with a missile – use it for destruction, or store it until it becomes obsolete and has to be scrapped. Either way, because the resources of the planet that went into making the weapon are fixed, 'We are now beating the ploughshares of future generations into present swords or warheads.'[127]

Education

Most of us have gone through the rather painful experience of cramming for an exam. The 'magic-marker syndrome' is a well-established academic phenomenon. That's when, the night before a test, the student takes out a yellow magic marker and proceeds to underline large sections of the textbook in hopes of memorizing and retaining giant chunks of data just long enough to regurgitate them back onto the test page in the classroom the following morning. Within twenty-four hours of the test, chances are good that little or none of the data has been retained. What has been retained, however, is a massive hangover which often lingers on for several days. Students get 'up' for the exam and afterwards they 'crash.' This is the typical pattern set in the American education system.

The way the student prepares for his exam is not unlike the way an ear of corn is prepared on an Iowa farm. In both instances, a massive expenditure of energy results in a slight entropy decrease in the product (in the student's case, the amount of knowledge retained) at the expense of a greater increase in the entropy of the environment. With the corn, the entropy increase in the overall environment is called environmental pollution. Psychologists now refer to the dissipated energy that accumulates in the student's environment as social pollution. It can manifest itself in a hundred and one different ways from the buildup of neurosis to nervous breakdowns.

Everything we do requires the expenditure of energy, even the learning process. The Entropy Law is always at work in the collection of information, as in every other endeavour. Of course, whenever we learn something, we generally believe that we are adding to the value and order of the world we live in. For a long time educators were convinced that at least the learning process was one activity that defied the second law by creating only greater order or the building up of nega-entropy. No longer. With the introduction of cybernetics and modern information theory after World War II, scientists realized that information gathering and the storage of knowledge required the expenditure of energy, and therefore an entropy price had to be paid.

Back some seventy years ago, Henry Adams wrote an essay in which he suggested that even the human mind, in its gathering and storing of information, was subject to the entropy process. The essay, entitled 'A Letter to American Teachers of History,' was addressed to the American History Association.[128] In it, Adams dared to suggest that the development in human thought over the ages had proceeded in the same direction as every other activity in the world; that is, toward a more and more complex, highly dissipating state. That essay caused quite a stir in academic circles at the time, for Adams committed the ultimate heresy. Like a thief in the night, he had stolen his way into civilization's inner temple, boldly flinging the second law across the most sacred altar of all: the one erected in honour of the spirit of the human mind. In the seven decades that have elapsed since he first put his thoughts to paper, Adams's essay has been rediscovered over and over again by academic scholars and made a subject of intense debate and discussion.

If Adams is to be accused of heresy then so should the ancient Greeks with their belief in the tale of Pandora's Box and the Jews and Christians with their belief in the account of Adam and Eve in the Garden of Eden. Both stories hold that the original perfection of the world was undermined with the introduction of knowledge. When Pandora lifted the lid to the box, opening up the secrets of life, and when Adam ate the apple from the tree of knowledge, it marked the beginning of a long and tortuous journey in which the accumulation and use of greater knowledge has led to greater disorder and fragmentation in the world.

Adams looked at the progression of the human mind – from instinct to intuition to reason to abstract mathematical thought – and concluded that each succeeding mental construct exhibited greater ordering, a higher energy flow-through, and, consequently, a greater dissipation of energy in the process. For example, comparing the instinctual responses of early man to his environment with the abstract rational responses of modern man to his environment, it is obvious, noted Adams, that in the former instance far fewer steps are involved in the mental process, and much less energy is dissipated.

In our own lives we sense that Adams's observation holds true. For example, we often hear people say that a gut reaction to a situation is more reliable than a reasoned decision. Or, that it's sometimes better to trust your own instincts in a given matter than your intellect. When asked why, the usual explanation given is that one's intuition or instinct is generally more closely attuned to the reality of what's occurring. That's true, and it has everything to do with the second law. As mentioned, the more stages in the mental process, the greater complexity, abstraction, and centralization, and the greater the dissipation of energy and disorder. The history of human mental development has been a history of removing the human mind farther and farther away from the reality of the world we live in.

The evidence also suggests that our mental activities have become more complex and abstract as our energy environments have become more harsh and exacting. After all, a hunter-gatherer environment required little more than raw instinct to survive in. Agricultural environments, on the other hand, require a great deal more abstract thought to manage. Industrial environments require even more.

A primary purpose of mental activity is to help the human being to survive. People survive by being able to locate and process available energy. As our energy environments have become more difficult to exploit, we have had to rely on a greater array of mental tools to order (and facilitate) our search and transformation activities.

It's also true that as humankind has developed its mental activity from instinctual response all the way to abstract mathematical reasoning, it has generated greater disorder in the world around it. The hunter-gatherers afflicted the world with far less damage than modern man and woman with our greater power of abstract reasoning.

The 'colonizing' period of human history has been characterized by the frantic depletion of one energy environment after another and the wreaking of greater and greater disorder on the earth. Still, the human mind continues to find new ways to collect, sort, store, and exploit greater amounts of information, in order to transform increased amounts of available energy through the system.

Today we are bombarded with information. Advertising, the mass media, our educational system are pounding on us with thousands and thousands of messages every day. From the time we get up in the morning until the moment we fall asleep at night, we are literally assaulted with bits of information. The advertising industry alone spent over $47 billion last year 'educating' the consumer.[129] The average American is subject to the one-way flow of information from the television set for over five hours every day.[130] Economic historians like Daniel Bell assert that our economy is now making the transition from an industrial to a postindustrial mode, where communication and information systems will dominate the economic activity of the nation.

This massive increase in information translates into a massive expenditure of energy. Along with it has come mounting disorders, increased centralization and specialization, and all of the other features that accompany a speedup of the entropy process. Already, the information and communication institutions – in both the private and public sectors – are turning into giant bureaucratic fiefdoms exerting enormous power over the lives of every American. The collection, exchange, and discarding of information is proliferating at an unparalleled speed. The increasing energy flow of the so-called information revolution is already creating massive disorders all along society's energy flow line, requiring more energy to be diverted into the ever increasing costs of maintaining the information and communication institutions and machinery.

The current computer and microchip 'revolution' is a case in point. Its advocates are fond of pointing to the fact that during the past thirty years, the prices for individual computers have plummeted dramatically, the size of the computer has decreased sharply, the amount of material resources as well as the energy necessary to run them has significantly dropped. At the same time, as computers have become smaller, cheaper, and less energy consumptive, the amount of information they can store, and the rate at which they can sift through facts, has increased astronomically.

Given all of this, it is easy to see why computer advocates argue that the computer is at least one example of how more and more can be done with less and less. The point should be self-evident, they say. After all, the day is near when the entire *Encyclopaedia Britannica* – all twenty-four volumes – can be stored on a single chip that costs just a few pennies. One day, it is forecast, you will have virtually all of the knowledge known to humanity at your fingertips, and you'll never have to leave your home to retrieve it. So, the computer actually uses less energy than traditional methods of accumulating information, makes it available faster, and opens it up for access to anyone who can afford the rather nominal computer purchase price. On the surface, this argument is rather convincing. However, notwithstanding these admittedly impressive points, the effect of the computer revolution, *in totality* has been to dramatically increase the overall entropy of the world. Any energy and resource savings evidenced by individual computers has been more than compensated for by the total entropy impact of computerization.

First, it's important to understand that while the average individual computer today consumes fewer resources and energy than the prototypes of thirty years ago, this very fact has led to an astounding proliferation of computers, whose numbers have necessitated a massive consumption of the world's resources. In 1950, just a few years after the birth of the first modern computers, only sixty computers had been built. But then, because of the transistor and the microchip, the full-fledged computer era was launched. By 1959, 6,000 computers were on line; by 1966, more than 15,000; by 1970, over 80,000, performing tasks in more than 3,000 different categories. Today, there are millions upon millions of computers, pervading every facet of life. IBM has already announced that it has more orders on hand for 1980 models than the total amount of all of its deliveries from the years 1950 to 1979! All of these computers use nonrenewable resources.

Second, it should be remembered that the computer is designed to collect, store, and disseminate information. The computer deals with facts, but those facts only take on real importance, in terms of entropic flow, when they are then used by society's technological transformers to collect, exchange, and discard energy. The computer is analogous to endosomatic sense organs. The mind uses the eyes, ears, and nose to see, hear, and smell; they are collectors of information. Yet, no animal could survive unless the sensory data thus collected are then used by its other endosomatic transformers – its legs, claws, teeth, and jaws – to collect and consume available energy from the surrounding environment. The more sophisticated the sensory apparatus, the better equipped an animal is to collect the information necessary to locate and convert available energy.

Similarly, the faster information is generated by the computerized society, the faster those sense data are used by the society's transformers to collect and convert available energy. The increased energy flow-through, in turn, creates greater disorder, a faster depletion of the existing energy base, and a greater concentration and centralization of the society's economic and political institutions. The very purpose, then, of the computer is to provide more sense data, more rapidly, in order to facilitate the faster conversion of available energy through the system.

It's also worth noting that as computers proliferate into every conceivable social function, society necessarily becomes dependent upon their workings for its survival. Computerization may make processes seemingly 'efficient,' but what is really taking place is that the computerized society becomes increasingly complex, and with complexity comes the real potential for breakdown. A single computer malfunction, for instance, can trip a series of switches in a major electric plant, shutting the facility down for days. Anyone who has ever gone

186

to an airline counter at a major airport when the computer is 'down' quickly experiences the frustration, and even helplessness, that comes from a computer malfunction. When the whole system becomes reliant upon a computer for its effective functioning, and when the single computer fails, the system fails with it. The human being becomes hostage to the technology.

Strangely enough, it seems that the more information that is made available to us, the less well informed we become. Decisions become harder to make, and our world appears more confusing than ever. Psychologists refer to this state of affairs as 'information overload,' a neat clinical phrase behind which sits the Entropy Law. As more and more information is beamed at us, less and less of it can be absorbed, retained, and exploited. The rest accumulates as dissipated energy or waste. The buildup of this dissipated energy is really just social pollution, and it takes its toll in the increase in mental disorders of all kinds, just as physical waste eats away at our physical well-being.

The sharp rise in mental illness in this country has paralleled the information revolution. That's not to suggest that the increase in mental illness is due solely to information overload. Other contributing factors include such things as genetics, spatial crowding, increased dislocation and migration of populations, and the stresses of environmental pollution. In less than twenty years, the concept of mental health in the United States has emerged from the academic halls to become a $15 billion industry. Today, upwards of 40 million Americans, or one out of five, are being treated for various mental illnesses.[131] As mental illness reaches what some health authorities consider epidemic proportions, a frenzied campaign is under way to set up appropriate treatment facilities. There are more mental health workers in the country now than there are policemen. They are part of a growing complex which includes

some 1,100 free-standing psychiatric outpatient clinics; 300 general hospitals with psychiatric outpatient services; 80 veterans hospitals with psychiatric outpatient clinics; 500 federally funded community mental health centres; tens of thousands of nursing homes, board and care facilities, halfway houses, behaviour clinics, child guidance clinics, child abuse, alcohol, and suicide prevention clinics.[132]

Not so long ago, Leopold Bellah, professor of psychiatry at New York University, compared mental health to public health, arguing that more needed to be done 'to protect the community against emotional contamination.'[133] While one might take exception to Dr Bellah's harsh language, there is no doubt that the term *emotional contamination* accurately describes what's taking place as we move more and more into an information and communication society.

Each of us experiences the effects of information overload every day – at work, in school, in the home, and out in the community. We find ourselves increasingly in the position of not wanting to know any more about a particular thing or about the world in general because we just can't handle it. Our nervous system and brain are only equipped to take in and use a certain amount of information at a time. When too much comes our way we attempt to filter out part or all of it by simply turning off. When incoming information zooms in on us from every direction, along with all kinds of fragmented background noise of every type and description, we experience extreme anxiety. How many times have you felt like you were about to either explode or be swallowed up by too many messages making demands on your attention?

Different people, of course, have different tolerances and thresholds. Everyone, however, has a limit beyond which the increasing flow of information and accumulated dissipation leads to breakdown and serious mental illness.

True to form, our society has devised a set of techniques to handle every imaginable human disorder, not realizing that the additional information infusion only alleviates one type of condition by bringing on other even worse ones. The image immediately comes to mind of the stereotyped therapy junky, a person who leapfrogs from one mental therapy programme to another in a desperate attempt to gain peace of mind, tranquillity, and an ordered life. By the time he's exhausted the entire smorgasbord of techniques and his own pocket-book, he is so overloaded with fragmented pieces of information, each with its own often conflicting prescription on how to 'cope' with the world, that he's a complete basket case.

Nowhere has the effect of the information revolution proven more deleterious than in our educational system. In the past fifteen years the cost of public education has quadrupled in the United States. In 1978, some 44 million children were being educated in the nation's public schools at a cost exceeding $81 billion.[134] Yet in the same period students have shown a steady decline in actual learning. As of 1979, over 15 per cent of all 17-year-olds in this country were functionally illiterate.[135] Many educators and parents are asking why kids are learning less when schools are equipped with more sophisticated teaching aids of all kinds, and a professional staff of teachers with specialized training in a host of academic fields. One woman who was interviewed in a CBS TV special on education summed up the apparent paradox. She pointed out that she had been educated in a tiny one-room schoolhouse in the South that contained some 'scratched-up desks, a raggedy book and a few crayons, and a few scratched-up blackboards.'[136] Still, she was perplexed as to why she was able to read and write but her children and their friends could not, even though their school had 'all the latest equipment.' Again, part of the answer is to be found in the speedup of the entropy process and the accumulation of disorders that follows.

Since World War II, the public school system has gone the way of many other institutions in American society. Smaller schools were absorbed into large centralized learning complexes. The uprooting of children from local neighbourhoods and the increased bureaucratization and specialization that went along with centralized educational institutions began to exact a toll in terms of student alienation, loss of discipline, and other disorders. Then, too, centralized educational complexes were able to make available all sorts of fancy new information technologies and specialized programmes to facilitate learning. All of these things combined have greatly increased the energy flow-through and resulted in problems ranging from increased learning disabilities to acts of vandalism and violence. Said one teacher on the CBS report:

189

We have this vast proliferation of distractors, diversions . . . We just keep pouring things into the school and on top of the kids' heads. And then suddenly someone realizes, 'Hey, this kid can't read.'[137]

The 'techniquing' of education has become so oppressive, remarked one parent in the same school system, that it's a wonder the children learn at all. The parent told of visiting the school's 'reading centre' and becoming depressed at how many instructions on 'exactly' how to read were plastered all over the room's walls. With that kind of overload, the parent remarked, 'I just don't see how they would ever really get into the mood of reading. If I were a child it would turn me off just totally.'[138]

The classrooms and corridors of America's giant school centres are overflowing with dissipated energy, much of it generated by the educational system. It's no wonder that children have a difficult time maintaining an attention span and that they exhibit a level of anxiety that often leads to outright violence. School vandalism now costs over $600 million a year.[139] Part of the blame, of course, rests with factors outside the school, many of which (but not all) are related to information overload. The TV is perhaps the number-one culprit. Five or more hours a day of nonstop one-way information flow is bound to seriously weaken the child's ability to concentrate and absorb information. Says one educator:

I think we've developed a generation that thinks of communication in terms of receiving messages, being talked to, rather than sending messages. And also, because of this great stimulation, when the child goes into the classroom they look at the teacher and, I think, make unhappy comparisons . . . It's not the excitement of a fleeting picture. Learning is hard work, and it cannot come with images on the screen.[140]

It's insane, when you stop just long enough to think about it, that as we become less and less able to deal with information overload, new techniques are devised by the media, the educational industry, and the information sciences to speed up, compress, and shove even more bits of information down us in hopes that enough will stay down long enough to be of some marginal economic or social value. Never once do they consider that the source of the increasing disorder rests with the very transformers that are directing the massive energy flow and increasing the entropy of the environment in the process. It reminds one of the story of the prison guards who were advised that increased punishment of the prisoners only increased their antisocial behaviour and the incidences of violence. After careful consideration, the guards concluded that the answer to the problem was to 'punish the increased violence.'

Health

Modern medicine, like almost every other activity in contemporary society, takes its cue from the Newtonian world view. The mechanical approach to medicine has dominated the health care profession for the past 200 years. British health expert Thomas McKeown sums up the prevalent attitude:

The approach to biology and medicine established during the seventeenth century was an engineering one based on a physical model. Nature was conceived in mechanistic terms, which led biology to the idea that a living organism could be regarded as a machine which might be taken apart and reassembled if its structure and function were fully understood. In medicine, the same concept led further to the belief that an understanding of disease processes and of the body's response to them would make it possible to intervene therapeutically, mainly by physical (surgery), chemical or electrical methods.[141]

Today, health care is the third-largest industry in the United States and accounts for nearly 9 per cent of the gross national product.[142] Much of the $150 billion ploughed into the medical field is for new, more complex, and more sophisticated technology gadgetry.[143] The modern clinic and hospital contain a plethora of diagnostic and therapeutic machinery. One of the major reasons behind the escalating costs of health care is the introduction of all of this medical hardware. The cost to the patient for this machinery is skyrocketing. Between 1950 and 1976, health costs per capita rose from $76 to $552.[144] Much of the increase went to pay for the enormous cost of maintaining the ever-expanding medical institutions. Today, the family doctor with a small individual practice has been eclipsed by the giant medical complex – centralized institutions housing hundreds of medical specialists and their machinery.

Centralization, increased specialization, and more elaborate equipment all translate into a greater expenditure of energy. As more energy has been expended in the medical field, the corresponding disorders have escalated. Although doctors don't like to talk about it, the sad truth is that the medical industry is no more immune from the Entropy Law than any other activity in society.

Chances are pretty good that you've never heard of the term *iatrogenic*, but every doctor has. Mention this little ten-letter word in front of a doctor and the response is likely to be one of instant defensiveness mixed with a slight tinge of terror. Iatrogenic diseases are those which are actually caused by the physicians, hospital, drugs, or machinery used to cure the patient.

The fact is, a temporary alleviation in condition following a medical procedure is often accompanied by an even greater long-range health problem for the patient. Part of the explanation for this lies in the fact that '75 to 80% of all patients seeking medical help have conditions that will clear up anyway or that cannot be improved even by the most potent of modern pharmaceuticals.'[145] Still, the doctors perform operations and prescribe various drugs, which create greater problems for the patients than the ones that sent them for medical help in the first place. For example, most of us are now aware that what little value (entropy decrease) we receive from having X-rays done is often more than outweighed by the long-range harm of radiation exposure (entropy increase).

We are also coming to understand how the entropy process works when it comes to the use of drugs. Every twenty-four to thirty-six hours, between 50 and 80 per cent of all adult Americans swallow a medically prescribed drug.[146] While they might experience a short temporary alleviation of their immediate discomfort or illness, the long-range deleterious effects of the drug on the human physiology are assured to be even greater. Nowhere is this more apparent than with antibiotics. These so-called wonder drugs are prescribed arbitrarily for just about every infectious disease that comes down the pike. The results have been catastrophic. Because antibiotics are indiscriminate killers of bacteria, they destroy many important organisms in the body that are absolutely essential to proper body maintenance. Vaginal thrush, yeast infections of the intestines, vitamin deficiencies, and a host of other disorders result from continued use of antibiotics. Then too, the massive use of these drugs has resulted in the proliferation of new resistant strains of bacteria, which have become so virulent that they can now survive both direct intervention and the normal healing activities of the human body. At an international symposium held on the subject in Linberg, West Germany, in 1976, many of the participants agreed that the human race is worse off than it would have been without the introduction of these so-called magic bullets.[147]

Antibiotics are only the tip of the iceberg. According to a detailed study published by a Senate subcommittee in 1962, of the 4,000 drug products legally marketed in the country over the past twenty-four years, almost half had no scientifically proven value.[148] Even more startling is the fact that many of these ineffective products, which are produced by major pharmaceutical houses, are actually dangerous and have caused ill health. In their book *Pills, Profits and Politics*, Milton Silverman, research pharmacologist, and Philip Lee, former assistant secretary of HEW, report that the adverse 'secondary disorders' caused by drugs 'kill more victims than does cancer of the breast.'[149] The problem has become so acute, say the authors, that adverse drug effects now 'rank among the top 10 causes of hospitalization and are held accountable for as many as 50 million hospital patient days a year.'[150]

It's impossible to know the full extent to which modern medical practices create even greater long-range medical disabilities. We do know, however, that even while patients are being administered treatment in a hospital, one out of five of them acquires an iatrogenic disease. One out of every thirty of these patients ends up dying from hospital-related illness.[151]

The tragedy is that for many patients there is absolutely no reason to be in the hospital in the first place. A congressional report found that, in 1974, doctors performed 2·4 million unnecessary operations, resulting in 11,900 unnecessary deaths, at an unnecessary cost to the public of $4 billion.[152]

Granted, the entropy process is at work. But the sceptic might well argue that modern medicine has at least been responsible for a 'temporary' improvement in the health and well-being of people, even if the penalty tax (the entropy increase) is beginning to come due. Statistics on the increase in life expectancy are most often conjured up as proof that modern medicine has produced some impressive results. This myth is tenaciously held onto because it provides society with the evidence it needs to continue to support a mechanical approach to medicine, and to life's other activities as well.

The reality is that modern therapeutic medicine has played virtually no role in the elimination of major death-causing illnesses and has little or no right to share in the credit for improvement in life expectancy. Several studies over the past few years have shown that the major contributing factors to improved life expectancy in the past 150 years have been better sanitation and hygiene and improved nutrition. One such study was conducted by John and Sonja McKinlay of Boston University and Massachusetts General Hospital. As in an earlier study done by McKeown in Europe, they found that the principal cause of the falling death rate in the United States since 1900 was the disappearance of eleven major infectious diseases: typhoid, smallpox, scarlet fever, measles, whooping cough, diphtheria, influenza, tuberculosis, pneumonia, diseases of the digestive system, and poliomyelitis. With the exception of influenza, whooping cough, and poliomyelitis, all of these infectious diseases declined almost entirely before medical intervention came on the scene. Overall, concludes the report:

Medical measures (both chemotherapeutic and prophylactic) appear to have contributed little to the overall decline in mortality in the US since about 1900 – having in many instances been introduced several decades after a marked decline had already set in and having no detectable influence in most instances.[153]

Up until 1950, the average life expectancy in America continued to climb. After 1950, it began to level off.[154] Today, for men at least, life expectancy has begun to drop. It's interesting that the retreat in life expectancy began to occur around the time that medicine began to take off into high-technology therapeutic health care. The 1950s also mark the early years of America's entry into the petrochemical age. On this last score, even the government now acknowledges a direct correlation between the rise in disease since 1950 and the pollution or high-entropy waste generated by our petrochemical economy:

The environment we have created may now be a major cause of death in the US. Cancer, heart and lung disease, accounting for 12% of deaths in 1900 and 38% in 1940, were the cause of 59% of all deaths in 1976 ... Growing evidence links much of the occurrence of these diseases ... to the nature of the environment.[155]

This is the conclusion of a top-level federal government task force composed of representatives of the EPA, the National Cancer Institute, the National Institute for Occupational Safety and Health, and the National Institute of Environmental Health Sciences.

The problem, according to the medical experts, is the tremendous rise in all forms of pollution at every level of human existence. In entropy terms, the high standard of living – the massive energy flow-through – we have enjoyed in this highly industrial environment is now being paid for with spreading disease and death. Pollution, let us recall, is simply the dissipated energy that accumulates from the energy flow of a society. The greater the energy flow, the greater the pollution, and eventually the greater the deaths that result.

The deadly effects of pollution on the human physiology are truly staggering. In New York City most taxi drivers have such a high level of carbon monoxide in their blood that it cannot be used for blood transfusions to persons with heart ailments.[156]

Recently scientists told a Senate subcommittee that it is no longer possible to find uncontaminated milk to feed to infants: 'Human breast milk increasingly contains pesticides, residues and other carcinogens. Infant formulas contain harmful lead deposits.'[157]

Several government reports in the past few years conclude that 60 to 90 per cent of all types of cancers in the United

196

States are caused by human-made environmental factors ranging from food preservatives and additives to toxic chemical substances.[158] Secretary of Health, Education and Welfare Joseph Califano shocked the nation's work force in late summer of 1978 by announcing the results of an extensive study showing that between 20 and 40 per cent of all cancers are work related – the result of contact with an entire range of metals, chemicals, and processes that are essential to the continuance of our industrial output. Because there is usually a time lag of from twenty to thirty years between exposure to chemical carcinogens and the onset of the cancer, it is estimated that as many as one out of every three Americans alive today will get cancer in his or her lifetime. In fact, since most of the sharp rise in industrial and commercial uses of synthetics, pesticides, and other chemical substances took place after World War II, many medical experts are predicting a virtual runaway epidemic of cancer by the mid-1980s.

Cancer is by no means the only major disease resulting from the pollutants of industrial society. The United Steelworkers union reports that 'more than a half million workers are disabled yearly by occupational diseases' of all kinds.[159] A study commissioned by the Environmental Protection Agency concluded that wages lost by American workers suffering from just air pollution alone total a whopping $36 billion per year.[160] Another study, done by the American Lung Association, estimates that the health bill for illness caused by air pollution totals over $10 billion per year.[161]

The health prospects for the immediate future are grim. Homo sapiens was not made for a highly industrialized petrochemical environment. Our anatomy has not changed since human beings first appeared on earth several million years ago. We were biologically designed for a hunter-gatherer existence. Each successive stage of economic and social development has only increased the physiological strains on the human being and further eroded our chances for long-range survival as a species.

Most diseases are environmentally induced. They are caused by the accumulation of waste (dissipated energy) as the entropy of a given environment increases. This isn't hard to understand. We all survive by drawing available energy from the environment. When the environment around us becomes choked and clogged with waste, it blocks the flow of available energy and pushes us closer to an equilibrium state.

Every energy environment creates its own unique form of dissipated energy. That dissipated energy, or waste, is internalized by different groups in society in different proportions depending upon how the energy flow line is set up. While it is true that throughout human history most of the major diseases have occurred within every kind of energy environment, the greater frequency of certain diseases over others can be accounted for by three related factors: the specific type of energy base of a civilization, the way the society's energy flow line has been set up, and the stage of the entropy process itself.

The argument that genetics plays more of a role than environment in the proliferation of certain illnesses is a bit of a misrepresentation. As René Dubos points out in his seminal work on the subject, *Man Adapting*, certain genotypes are less resistant than others to a particular environmental disease and therefore more likely to be affected. But still, it's the nature of the energy base, its state of entropy, and the way the energy flow line is set up that determine the likelihood of specific disease epidemics. For example, infectious diseases were virtually unknown in the hunter-gatherer environment, where the communities were small, extremely mobile, and lived an outdoor existence.

In agricultural environments, where there is a close living relationship between the sedentary populations, domesticated animals, small rodents, and so on, microbial agents are the main source of disease. As the energy environments become more depleted by the expansion of agricultural lands, the felling of forests, and the erosion of soil, natural habitats become more and more disrupted, allowing for the spread of certain microbial agents. The kinds of infectious diseases that occur are a result of the types of imbalance created by the energy flow lines.

In the advanced industrial environment, the chief cause of disease is the dissipated energy created by our energy base of

198

nonrenewable resources. As already mentioned, the increased frequency of cancer, heart disease, and other chronic and degenerative illnesses is more and more found to have a direct relationship with the nonrenewable energy base. The proliferation of these diseases, in turn, follows directly on the heels of the increase in the entropy of the environment. Finally, the frequency of these types of illnesses is found to vary from group to group in the population depending upon their position in the flow line – that is, the type of work they do, the amount of energy (income) they are compensated with, the places where they live their lives, and the type of life-style they adopt.

As the dissipated waste created by our high flow-through nonrenewable energy sources continues to build up all along society's energy flow line, causing a dramatic escalation in physical disorders of all kinds, a point will be reached where the population will have no choice but to shift back into a low flow-through renewable energy base or face disease and death in epidemic proportions.

PART SIX

Entropy:
A New World View

Toward a New Economic Theory

There is no easy way to make a transition from a mechanical world view based on the idea of permanent material growth to an entropic world view based on the idea of conserving finite resources. But let there be no mistake about the consequences of holding on to the traditional way of doing things. In 1979 there were scattered reports of shootings at petrol lines in various parts of the United States. Several people were murdered over a gallon or two of petrol. Life is going to get a lot uglier in the years ahead. We're running out of nonrenewable energy, and unless we come to grips with that realization and begin making the radical readjustments that are necessary, we will see more blood spilled in American streets. It is just as likely that the worsening energy crisis will lead to American military intervention in the Middle East, an all-out nuclear confrontation, and holocaust for the planet. In 1979, the United States Army announced that it was 'drafting plans for a new "quick strike" force of 110,000 troops to respond to crisis in the Persian Gulf.'[1]

These observations are not made lightly. Never before in American history have we faced such a grave threat to our way of life. If the flow of nonrenewable energy slows sufficiently to grind the American economic machine to a standstill, the hue and cry for immediate action will be deafening. There will be no more liberals or conservatives then, no more hawks or doves. Only millions of desperate people seeking relief at any cost. That time is not way off in the distant future. It could come at any moment.

The alternative to panic and bloodshed is a difficult one, the most difficult any civilization has ever had to undertake since the beginning of history. It took thousands of years to make the transition from a hunter-gatherer existence to an

203

agricultural one. It took hundreds of years to move from an agricultural way of life to an industrial one. In both instances there was plenty of time to make the radical adjustment in world views that was necessary to accommodate the new economic circumstances. Today we are being forced to make a transition from the Industrial Age of nonrenewable resources to a new and still undefined age based once again on renewable sources of energy, and we will have to do so in little more than one generation. The radical change in world view required to make this transition will have to be accomplished virtually overnight. There will be no time for polite debate, subtle compromise, or momentary equivocation. To succeed will require a zealous determination – a militancy, if you will – of herculean proportions.

When we hear proponents of solar energy espouse the great benefits of moving our energy base from nonrenewables to sun power, the impression conveyed is that the whole transition can be accomplished without revolutionary changes in our way of life. This just isn't the case. Different technologies and institutions are designed for different energy environments. The types of transformers that will make up the Solar Age will be completely different from those we now live with in the age of fossil fuels.

What must be understood is that the Industrial Age is nothing more than a name for the kind of transformers that have been established in response to the nonrenewable energy base we have lived off. While there have been socialist and capitalist flow lines, all industrial countries exist only by the grace of the nonrenewable energy base upon which their economies depend. *The end of the age of nonrenewable energy, then, presages the end of the Industrial Age as well.* As the stored nonrenewable energy runs out, the entire economic superstructure built upon it will begin to crumble. Cracks all along that superstructure are already beginning to appear, and, try as we will, there isn't enough nonrenewable energy left to mend all of them. This is the hard truth that every person on this planet must ultimately face up to.

In the long run, the Solar Age we are moving into will function as differently from our Industrial Age as we have functioned differently from the medieval era that preceded us. In the short run, several key transition steps are essential if we are to make the successful transformation from our age to the next.

Third World Development

According to the prevailing wisdom, the more the industrial economies grow, the more the rest of the world will benefit. This assumption rests on the idea that the faster we can convert raw resources into economic goods, the more permanent value or wealth we create that can be divided among the peoples of the earth. With this as the central principle of international economic development, it's no wonder that advances in technology are viewed as the source for creating more and more 'permanent' wealth. The laws of thermodynamics, however, provide a very different frame of reference. The fact is, the faster the developed nations convert raw resources into economic goods, the less is available in nature's storehouse for other countries and for future generations. Advances in technology, for the most part, serve to speed up the conversion of more resources in a shorter period of time depleting nature's stock and creating even greater waste and disorder in the process.

As this reality begins to sink in, it's important to recognize the fact that the remaining reservoirs of untapped nonrenewable resources are primarily in the hands of the poor Third World nations. These resources are their only remaining trump card to bargain for a more equitable redistribution of wealth between the industrialized countries and their own. The Middle East oil-producing countries are already using that leverage effectively. Their cartel arrangement to control the terms and the flow of oil exports is now being copied by other Third World countries dealing in other nonrenewable resources. Cartels have now been established to regulate the price of bauxite, copper, iron, chromium, and lead. Says *Fortune* magazine, 'If the material exporters succeed in this endeavour, the days of sustained improvement in living standards in the advanced industrial countries may well come to an end.'[2]

To those of us who have lived for decades on huge quantities of energy and resources provided by the Third World, it is easy to resent the squeeze that cartels will put on our economic system. A popular country-and-western song of the summer of 1979 summed up the frustration many Americans felt over escalating OPEC oil prices: 'No crude, no food.' In other words, if the Third World won't sell us its petroleum, then we should withhold food exports from the world's hungry. This kind of jingoistic attitude on our part is not only morally and politically indefensible, but it threatens our very survival. The choice is ours. We can either accept the new terms presented by Third World nations and cut back dramatically on our energy flow and material consumption, or we can intervene militarily to seize the resources we need. This latter choice would not go unresisted by the Soviet Union and other world powers. Are we so hooked on cheap resources taken from other countries that we are willing to start a process of aggression that could quickly lead us into a Third World War and the obliteration of the planet by nuclear weapons?

Most of us simply do not understand what is taking place in the Third World. While we pay lip service to the tragedy of squalor, hunger, and overpopulation in the Southern Hemisphere of the globe, we really have no conception of the misery in which over half the planet lives. Fully 800 million human beings are barely surviving in what the World Bank calls 'absolute poverty' on an annual income of $200 or less.[3] Fifteen to twenty million Third World deaths annually – three-fourths of them children – directly result from malnutrition. As you read this, twenty-eight people are dying in our world each minute as a consequence of hunger.[4] Eighty per cent of the world's population, mostly living in villages, have no system of health care.[5]

We Americans shake our heads in sadness at this incredible dehumanization. Yet the plain fact is that as long as we in the United States continue to consume one-third of the world's resources annually, the Third World can never rise to even a semblance of a standard of living that can adequately support human life with dignity. Those who are irate over the formation of resource cartels as an economic weapon to be used against rich nations like our own had best ask themselves what they would do if they were living in the Third World. Any Third World leader who says he would continue to allow the industrialized nations to plunder his country's natural resources is a fool.

To Third World people, talk of the era of limits, curbing material expectations, establishing no-growth economic policies, and the like appear to be just one more attempt by the industrialized nations to keep poorer countries in their place of international subservience. Third World countries, just beginning their own industrial output, see the ecological concerns of the rich as little more than an effort by countries like the United States to hold on to their wealth by hindering economic growth among the poor. In a paper presented before the World Council of Churches' 1979 Conference on Faith, Science and the Future, C. T. Kurien spoke for many when he offered a Third World perspective on the 'limits to growth' thesis:

It is a small affluent minority of the world's population that whips up a hysteria about the finite resources of the world and pleads for a conservationist ethic in the interests of those yet to be born; it is the same group that makes an organized effort to prevent those who happen to be outside the gates of their affluence from coming to have even a tolerable level of living. It does not call for a divine's insight to see what the real intentions are.[6]

Kurien's point is well made. As long as we continue to devour the lion's share of the world's resources, squandering the great bulk of them on trivialities while the rest of the world struggles to find its next meal, we have no right to lecture other peoples on how to conduct their economic development. Therefore, if we are truly committed to preventing our planet from being turned into a giant industrial sewer, we must begin, now, voluntarily, to substantially limit our own material wealth. We must show our own willingness to accept hard sacrifices in the name of humanity.

However, this too must be said: no Third World nation should harbour hopes that it can ever reach the material abundance that has existed in America over the past few decades. To put its faith in Western-style development is a cruel hoax, simply because it is a physical impossibility even if there were a complete redistribution of the world's resources. According to economist Herman Daly:

If it requires roughly one-third of the world's annual production of mineral resources to support that 6% of the world's population residing in the US at the standard of consumption to which it is thought that the rest of the world aspires, then it follows that present resource flows would allow the extension of the US standard to at most 18% of the world's population, with nothing left over for the other 82%. Without the services of the poor 82%, the 'rich' 18% could not possibly maintain their wealth. A considerable share of world resources must be devoted to maintaining the poor 82% at at least subsistence. Consequently even the 18% figure is an overestimate.[7]

It is thus impossible for the rest of the world to develop as the United States has. In fact, as we have already seen, absolute resource scarcity makes it impossible that even the United States can continue at anything near its present level of energy flow. This is not, however, to dismiss the absolute necessity of fostering economic development in the Third World. The question is: What kind of development is appropriate to poor nations?

Unfortunately, many Third World countries are using their new-found wealth to industrialize their economies along the same line as the United States and other so-called developed nations. Their ill-conceived economic policies can only lead to tragedy for both their own nations and the planet as the entropy process escalates even faster toward a watershed. First of all, at a time when the world is running short of nonrenewable resources it is foolish to develop an economic infrastructure based on a high energy flow of nonrenewable resources. Third World nations like Brazil and Nigeria will have built a massive industrial infrastructure by the year 2000, only to find that they can no longer secure adequate amounts of nonrenewable energy to keep the economic machinery running.

When Western-style progress comes to a Third World

nation, 'instant underdevelopment' is usually the result. In other words, the masses of people of these countries actually become poorer than before development began. The major reason for this is that Western industrialism favours cities over rural areas and highly centralized, energy- and capital-intensive production over human labour. As nations seek to industrialize, jobs actually diminish because production is automated. At the same time, mechanized agriculture promoted by the much-vaunted Green Revolution has the effect of forcing peasant farmers off the land. This is so because mechanized agriculture requires expensive inputs of energy into the farming process. Because of this, small farmers are squeezed out of the market. Dislocated peasant farmers are forced to move to the city to try to find jobs. This process is taking place all over the Third World. By the year 2000, it is estimated that 1 billion more people will be jammed into Third World urban areas than lived there in 1975.[8] As forced urbanization proceeds, greater poverty ensues. Further, as agriculture around the world follows the US model, the world's food situation becomes more precarious because farming becomes increasingly reliant on nonrenewable resources. If the entire world converted to our style of agriculture, up to 80 per cent of all energy conversion would go into food production, and we would exhaust all petrochemicals within a decade.[9]

High-energy industrial development also brings with it other disruptions to traditional patterns of living. In the 1880s, it is said, a Saudi Arabian sheikh discovered oil bubbling out of the sand in a remote desert. He ordered the hole filled in, and forbade anyone to reveal what they had seen. Why? Because he feared that Westerners would come barrelling in with their technology and contempt for tradition. The sheikh's motives might well be suspect, but he was certainly correct in his fears. As high-energy technology is exported to the Third World, it brings with it a certain ideology. Third World leaders continue to naively assume that they can bring in the wealth and technique of a country like the United States but not bring with it a set of modern technological values that are destructive to the traditional culture.

It is clear that Third World nations must seek different forms of development from those used in the industrialized West. High-energy, centralized technology should be eschewed in favour of intermediate technology that is labour intensive and can be used in local villages – instead of necessitating the mass migration of people from rural communities to squalid, overcrowded cities. Agriculture will still have to form the base of Third World societies. Because of its current development pattern, Arab nations now import 50 per cent of their food. By the year 2000, they will be importing 75 per cent.[10] For these and other Third World countries any sound policy of development should refocus on the establishment of a labour-intensive farming base that can provide a society with a self-sustaining food source.

Several appropriate models for Third World development already exist. Before Mao's death, the People's Republic of China organized itself in a way that maintained the rural base of the society and favoured labour-intensive production. China is not a rich society, but very few people are jobless or homeless. More attention should also be turned to the Gandhian economic model. During the anticolonial movement led by Gandhi, the symbol of the struggle became the hand-operated spinning wheel, a simple piece of appropriate technology that allowed each Indian to have some control over his or her own economic livelihood in even the poorest or the most remote village. Gandhian economics favours the country over the city, agriculture over industry, small-scale techniques over high technology. Only this general set of economic priorities can lead to successful Third World development. But once again, it must be said that high-energy-flow nations like the United States must be willing to undertake sacrifices.

Domestic Redistribution of Wealth

Accepting higher and higher prices for all nonrenewable resources means a steady contracting of the American economy. For the first time in our country's history we will have to deal with the ultimate political and economic question – redistribution of wealth. In the past this question has always remained on the periphery of our national agenda. As long as the economy was continuing to expand, there were always enough marginal gains – or leftovers – to pacify or buy off those at the bottom of the economic pyramid. Now that the economy is contracting, the call for redistribution of the remaining share will be heard from many quarters: not only the poor, but the working class and middle class as well are likely to join in demands to redistribute both wealth and power.

Today, the top one-fifth of the American population consumes over 40 per cent of the nation's income.[11] This is also the class that exercises control over the institutional machinery, that is, the energy flow line of the nation. The battle between this class and the poor is going to be vicious; the outcome is likely to hinge on how successful each of these two groups is in recruiting the large middle-income sector to its side.

The contraction of the American economy has already begun. On September 6, 1979, the Secretary of the Treasury warned the nation that it must go through 'a period of austerity.' Austerity, he noted, will fall most heavily on the poor, the elderly, and those on fixed incomes.[12] While the Secretary said that ways could be found to make monetary transfers to these people to lessen their burden, there is really only one viable solution: it is imperative that there be a massive redistribution of wealth and power in this society. Without that redistribution, the poor and working classes in America will rightly condemn any talk of austerity or econ-

212

omic sacrifice in much the same way as Third World nations inveigh against the wealthy countries preaching the gospel of limits.

In nature, whenever one element of an ecosystem multiplies or grows out of proportion to its proper functioning relationship with the rest of the elements in the system, it robs other life forms of the negative entropy (available energy) they need to survive. By doing so, it threatens the continued existence of the entire system. This is also the case in human society. When certain individuals or institutions capture an inordinate amount of the society's energy for themselves, their gross accumulation of wealth and power robs the rest of the members of society of the available energy they need to survive. History shows that whenever a society's energy (wealth) becomes so concentrated in the hands of a few individuals or institutions that the rest of the society suffers energy deprivation so great as to imperil their own survival, the society either crumbles or moves to revolution or both. While nature relies on self-regulating biological laws to restore balance, society must rely on agreed-upon principles of economic justice to achieve the same ends.

Slowing down the entropy process to a pace more closely attuned to the entropy process in nature requires both minimizing energy flow-through and redistributing the smaller amounts of energy more equitably among all members of society. Unless both are done at the same time, it is unlikely that the social order could survive intact during the transition period to a new energy base.

Without a fundamental redistribution of wealth, all talk of lowering energy flow and heeding our planet's biological limits will result in nothing but the rich locking the poor forever into their subservient status. The chic upper-class ecologists, with their hot-tubs, their quarter-million-dollar homes, their designer clothes, and their Mercedes Benzes, had best realize that their calls for clean air must be accompanied by meaningful actions that will lead to a redistribution of their own unwarranted economic abundance. If they do not voluntarily begin to make this economic adjustment, then others will make it for them.

213

A New Infrastructure for the Solar Age

All of us know the old cliché 'You can't have your cake and eat it too.' But most of us prefer to believe just the opposite. Such is the case now, as our society begins to herald the virtues of solar power. There is no doubt of the superiority of solar-based power over any other form of energy. Moreover, economic realities guarantee that the world will move inevitably toward a Solar Age. Still, few people have stopped to analyse the profound implications of this shift in the energy base. Many seem to think that the Solar Age will be just like our own, only cleaner. Cars will be electric, so there will be no smog; cities will run on solar collectors; homes will be inexpensively heated and cooled through sun power; solid waste will be converted at biomass plants into gasohol; quaint windmills will dot the landscape, reminding us of a more tranquil past, even as the pollution-free machines of industry churn on, quietly manufacturing consumer goods to sustain our style of living. In the Solar Age, it would seem, you *can* have your cake and eat it too.

In fact, this is far from the truth. The transition period to the Solar Age will require a complete reformulation of economic activity at every level of American society. Once we grasp the enormous implications of shifting the energy base of society from a concentrated stock (fossil fuels) to a diffuse flow (solar), it becomes apparent that our existing industrial structure is completely unsuited to a solar future.

In the most literal sense, highly concentrated nonrenewable energy has shaped today's economy. In order to maintain the existing institutional superstructure, we would have to continue to rely on a highly concentrated flow of energy through the system. Solar energy, however, is not concentrated like nonrenewable energy and is therefore unsuited for a highly centralized industrial life-style.

While there are many different techniques for harnessing

214

energy from the sun – solar thermal projects, photovoltaics, wind power, biomass conversion – and numerous modes of collection, ranging from sophisticated high-tech systems to ancient passive solar systems, all are based on tapping into a diffused flow of energy, rather than a concentrated stock. As a flow, solar power has the obvious advantages of being clean, abundant, and virtually inexhaustible (that is, until the sun burns itself out in billions of years). At the same time, there are inherent disadvantages, at least in terms of trying to maintain our contemporary form of social existence.[13]

Because solar radiation is diffuse, it must be concentrated to do work. Since the laws of thermodynamics tell us that work can only be performed when there is a temperature difference between two places, and since solar energy falls essentially equally on each square foot of land in any given geographical area, the solar flow must be collected. If electricity is desired, the stored solar energy must be transformed from one state into another. The nature of the flow and the economies of scale of solar technologies are most suited to small units, such as those that could provide enough heat and hot water for an individual home. Most solar advocates agree that at this stage of the technology, and for the foreseeable future, converting existing private homes to solar power will only provide 60 per cent of the dwelling's energy needs.[14] While far-more-efficient solar homes could be built from scratch, the conversion will be a slow process; 75 per cent of all the structures that will exist in the United States by the year 2000 have already been built.[15]

At the industrial and urban level, solar energy does not lend itself at all to the complex technological organization required by contemporary society. One estimate, for example, indicates that in order to run our current industrial superstructure we would need to cover between 10 and 20 per cent of the total US land area with various types of solar collectors.[16] Another estimate shows that Manhattan daily consumes more than six times the energy that could be provided by a 100-percent-efficient collector of all the solar flow that falls on the city.[17] To power New York City through various solar techniques, an area many times the city's mass would have to be given over entirely to solar collectors. While New York is obviously unique in its consumption of energy, other major urban areas will be subject to similar strictures in the Solar Age.

The sheer size of the solar infrastructure that would have to be erected to maintain modern society is mind-boggling. So too is the amount of time and labour that would be required to build and sustain it. For example, to equip just 3 million houses with solar collectors would require a work force of 200,000 people producing and installing 800 million square feet of solar collectors at a cost of $20 billion.[18] To construct a solar energy base for a major urban area would take millions of workers. As E. F. Schumacher wryly noted, 'While you can heat a house with solar energy very comfortably, you can't heat Rockefeller Center. In fact, solar energy plus wind power would not push the lifts up and down. And most of the accommodations in Rockefeller Center are inaccessible if there are no lifts: Fancy someone climbing thirty or fifty floors.'[19] Schumacher contends that large-scale production and urban living do not fit the model of the Solar Age. Murray Bookchin, the author and ecologist-anarchist, agrees. Bookchin argues that solar power and wind power 'cannot supply man with the bulky quantities of raw materials and the large blocks of energy needed to sustain densely concentrated populations and highly centralized industries . . . Solar devices . . . will produce relatively small quantities of power.'[20] Ecologist William Ophuls concurs: 'Conversion to exclusive dependence on solar energy would clearly require major changes in our technology and economy in the direction of greater frugality and decentralization.'[21]

These statements seem to make common sense when viewed in the context of thermodynamic realities. Still, to some they appear heretical. Many solar advocates, for instance, claim that frugality or austerity should not be equated with a sun-powered future. All the energy we require will be available in the coming age, they say. The fallacy in this argument is an unstated myth that solar energy performs work on its own, and since it is non-polluting and renewable, the more solar power moving through the energy flow lines, the better.

Though the evidence suggests otherwise, let us assume just for the sake of argument that new collecting techniques could be discovered that would allow us to very efficiently concentrate the flow of solar radiation far beyond the capacity now available or even deemed conceivable by many engineers. If this remarkably efficient recovery process were somehow possible, we could then support an urbanized, industrial-technological society through solar flow. But what would be the result? Simply this: we would continue to witness the exponential increase of entropy here on earth as solar energy is used to convert more and more of our limited terrestrial energy resources (matter) into the production process, transforming them from a usable to an unusable state. It is not, then, just the form of energy a society uses that is critical; it is also the amount of energy. If solar energy actually could flow in highly concentrated forms for industrial use, we would experience many of the same economic and social dislocations that result from our high energy use now. That's because *the use of solar energy cannot be divorced from the stock of fixed terrestrial matter that it interacts on and converts. In living and in industrial processes, solar energy must always be combined with other terrestrial resources in order to produce a product. That conversion process always results in the further dissipation of the fixed stock of terrestrial resources on the planet.*

There are many current proposals for diverting as much of the remaining flow of nonrenewable energy as possible into the building of a new infrastructure of institutional and mechanical transformers that would collect, store, process, and direct solar energy. While many of these proposals make sense, it should be made abundantly clear that any such infrastructure is only temporary and can serve as little more than a tiny halfway house to soften the impact of transition. In the long run, a solar infrastructure derived from and dependent on nonrenewable resources cannot be supported on a scale necessary to maintain highly industrialized economies. The nonrenewable resources will simply not be available in the quantities required.

In examining the potential of solar power, environmentalist Howard Odum has developed the concept of *net energy*. Net energy is the energy yield of a technology, minus the energy invested in recovering it. Odum argues that 'solar energy can generate some net, concentrated energy in the form of food, fibre and electricity, but the amount per area is small because most of the solar energy is consumed by the various structures that have to be maintained and operated to collect and concentrate the energy.'[22]

While Odum is clearly in favour of solar use over coal, uranium, or oil, we must realize, he says, that solar technologies will require vast amounts of nonrenewable energy and material resources to make the millions upon millions of solar devices that will be required. In essence, it will be necessary to construct an entirely new energy infrastructure for society. Though solar techniques may not be as capital intensive as petroleum refineries and synfuel plants, the volume of scarce resources required is awesome. Converting just 2·5 million homes to 60 per cent solar efficiency, for example, would consume fully one-third of all US copper production.[23] If one-half of all US electricity were produced through solar fuel cells – the most efficient converters now available – the building effort would annually require more platinum than is yearly produced worldwide.[24] Truly huge amounts of other nonrenewable resources would also be necessary to build a massive solar infrastructure. Among these are cadmium, silicon, germanium, selenium, gallium, arsenic, and sulphur; as well as megatons of glass, plastics, and rubber; and great volumes of ethylene glycol, liquid metals, and Freon. According to one source, 'If we settle on cadmium-sulphide cells for direct photovoltaic conversion ... it would require the entire 1978 world population of cadmium to produce only 180,000 megawatts of installed capacity, or about 10 per cent of the capacity the world had in place last year.'[25]

Writing in the *Atlantic Economic Journal* in December 1978, Nicholas Georgescu-Roegen pointed out the obvious flaw in current approaches to the harnessing of solar energy:

The truth is that any presently feasible recipe for the direct use of solar energy is a 'parasite,' as it were, of the current technology,

218

based mainly on fossil fuels. All the necessary equipment (including the collectors) are produced by recipes based on sources of energy other than the sun's. And it goes without saying that, like all parasites, any solar technology based on the present feasible recipes would subsist only as long as its 'host' survives ... The intensity of solar radiation reaching the ground level being extremely weak, a large material scaffold is needed for its collection ... It is highly plausible that the difficulty may not be superable at all, given that the intensity of solar radiation is a cosmological constant beyond our control.[26]

Our future is a solar future, of that there can be no doubt. The question is whether we will continue in our old habit of thinking and futilely attempt to generate a high-technology, resource-intensive solar energy base that will hasten the degradation of the planet, or whether we will generate an energy base that, at every step of its formation and use, seeks to keep the flow of energy and resources at a minimum.

Not surprisingly, the high-tech, resource-intensive mode is favoured by big business. Of the nine largest photovoltaics firms (photovoltaics are collectors that store sunlight in batteries for later use as electricity), eight are now owned by large corporations, five of them major oil companies. According to Richard Munson of the Solar Lobby, Exxon and ARCO will soon control more than half the industry between them. Other facets of solar technology are also being gobbled up by big business. For example, twelve of the top twenty-five solar companies are now controlled by corporate giants with annual sales of $1 billion or more. Among them: General Electric, General Motors, Alcoa, and Grumman. Obviously, it is the goal of these companies to ensure that solar power is developed in as high-technology and centralized a manner as possible.[27]

The 'big is beautiful' solar strategy is already leaving the drawing board and moving into actual production. Aerospace firms, for instance, are lobbying heavily to induce the government to fund 'Sunsat,' a solar satellite that will be bigger than the island of Manhattan. And in Barstow, California, McDonnell-Douglas, backed by hefty federal funding, is completing work on its 'power tower.'[28] The $130 million project consists of 2,200 giant mirrors that will focus sunlight on a boiler atop a 500-foot concrete tower.[29] These schemes are obviously solar technologies developed by a fossil-fuel mentality; that is, they attempt to concentrate a diffuse solar flow as much as possible in the hopes of turning it into a centralized energy stock, much like coal and oil. The attempt, however, will only cause greater disorder than any possible value gained. The amount of nonrenewable energy resources that go into constructing the parts of a giant solar satellite and launching it into space where it must be assembled is far greater than the amount of energy the Sunsat could produce for many years. Concentrating solar rays in such high density and beaming them back to collectors on earth will cause microwave radiation pollution that will endanger the health of anyone living or working near the collector. Some areas of the country might be deemed uninhabitable because of the microwave danger. Once the power has been collected at a central location, it must then be shipped as electricity through power lines. This will require the use of more quantities of nonrenewable resources to construct this part of the infra-structure. The 'power tower' suffers similar problems. The more concentrated the collection of solar rays, the less net energy will remain.

Even on a smaller scale, there are important choices to be made. For example, in home units, solar power can be provided through either low or high technology. The higher the technology used, the less net energy will be provided, because the more nonrenewable resources must be used to build and maintain the collecting infrastructure. For instance, in the high-tech – or *active* – home system, sunlight is first concentrated in a collector made of nonrenewable resources; then the solar energy is stored in either air or water housed in containers manufactured of nonrenewables; finally it is moved by fans or pumps to perform the work required. Another high-tech system is one in which photovoltaics concentrate energy and store it in batteries. Once again, nonrenewable materials form the base for the technology. While these systems clearly use the less intensive form of technology than do solar satellites and power towers, small-scale home units

220

of an active nature must still depend ultimately on the supply of copper, platinum, and other diminishing ores out of which the solar utilization equipment is manufactured.

Passive home solar systems tend to be less ecologically damaging and provide the most net energy yield because they are based less on nonrenewable technology and more on the life experiences of the first Solar Age that preceded the fossil fuels era. In a passive system, homes are actually designed and constructed in such a way that they naturally remain cool in summer and warm in winter. Although many workable prototypes of passive solar homes have recently been developed by architects, anthropologists can point to workable systems that were developed hundreds and even thousands of years ago by peoples who had no other way to maintain their homes.

The Solar Age will require a greater conformity to the ancient rhythms of life. While small, appropriate technology relying on very limited remaining stocks of nonrenewable energy will still be used where absolutely essential, the bulk of the transforming work will revert back to human and animal labour as it has in every other period of history before the Industrial Age.

Those wedded to the Newtonian world view and the Industrial Age will no doubt regard these observations about solar technology as pessimistic. Many will consider it inconceivable that urban life, industrial production, and all the creature comforts that make up the so-called American Dream are antithetical to the Solar Age. However, ecologists and economists like Georgescu-Roegen, Daly, Odum, Bookchin, and Ophuls would argue that to ignore the historical reality in front of us in favour of maintaining false expectations is sheer madness and will lead to an even greater fall for humankind, perhaps an irreversible one. Regardless of which course we follow, the coming transition is sure to be accompanied by suffering and sacrifice. But there is really no other choice. The fact is, the suffering will be minimized if the transition from the existing energy base to the new one is made now in a thoughtful, orderly manner, rather than later, out of sheer panic and desperation. We are

221

rapidly approaching the absolute limits of the fossil fuel energy environment. If we wait until we run smack up against the wall of this existing energy base, we will find that we have no energy cushion left to ease the transition process.

Values and Institutions in an Entropic Society

In a high-entropy culture, the overriding purpose of life becomes one of using high energy flow to create material abundance and satisfy every conceivable human desire. Human liberation is thus equated with the accumulation of greater wealth. A primary value is placed upon transforming the environment to extract its riches.

Having banished God from society, the high-entropy, materialist value system attempts to provide a heaven on earth. In so doing we have placed man and woman at the centre of our universe, and defined the ultimate purpose of our existence as the satisfaction of all possible material wants, however frivolous. We have reduced 'reality' to that which can be measured, quantified, and tested. We have denied the qualitative, the spiritual, the metaphysical. We have entered into a pervasive dualism – our minds separated from our bodies, our bodies divorced from the 'surrounding' world. We have gloried in the concepts of material progress, efficiency, and specialization above all other values. In the process, we have destroyed family, community, tradition. We have left behind all absolutes, except for our absolute faith in our ability to overcome all limits to our physical activity.

Now our world view and social system are falling victim to the very process of their creation. Everywhere we look, the entropy of our world is reaching staggering proportions. We have become creatures struggling to maintain ourselves in the midst of growing chaos. Each day we experience the truth that biologists have long known: an organism cannot long survive in a medium of its own waste.

There is no doubt that we are in for a massive institutional realignment. Our social structure, geared as it is for a maximum energy flow, will no longer be sustainable. Our institutions – their configuration, their purpose, their method of operation – will be radically transformed. But before we

223

can even begin to broadly outline the nature of agriculture, industry, and commerce in a low-entropy society, we must turn our attention to first principles, those underlying values that give meaning and direction to our lives.

On his 1977 lecture tour of the United States, E. F. Schumacher noted: 'The most urgent need of our time is and remains the need for metaphysical reconstruction, a supreme effort to bring clarity into our deepest convictions with regard to the questions What is Man? Where does he come from? and What is the purpose of Life?'[30] These are the Big Questions of human existence, questions which have absorbed people for thousands of years. Today, in our nine-to-five existence, they are not much discussed and are, in fact, dismissed as being 'prescientific' because they do not fit into the neat little standardized explanations of the world offered us by the Newtonian world view. Nonetheless, the Big Questions of the past are destined to reemerge in the low-entropy world that awaits us. For a low-entropy energy environment provides a completely different orientation to the goals of humanity. The governing ethical principle of a low-entropy world view is to minimize energy flow. Excessive material wealth is recognized as an irreversible diminution of the world's precious resources. In the low-entropy society 'less is more' becomes not a throwaway phrase but a truth of the highest magnitude. A low-entropy society deemphasizes material consumption. Frugality becomes the watchword. Human needs are met, but whimsical, self-indulgent desires – the kind pandered to in every shopping centre in the country – are not.

The traditional wisdom, as embodied in all the great world religions, has long taught that the ultimate purpose of human life is not the satisfaction of all material desires, but rather the experience of liberation that comes from becoming one with the metaphysical unity of the universe. The goal is to find 'the truth that will set us free'; to find out who we really are; to identify with the Absolute Principle that binds together all of existence; to know God. In Sanskrit, it is put most succinctly: *Tat tvam asi* (That art thou). To know this in the very ground of our being and to conduct our life in accordance with this

224

transcendent reality: this is the human development that comes from an adherence to traditional wisdom.

Unwarranted consumption, possessions, and a general attachment to material things have been discouraged by all the great religious teachers of the past:

The cultivation and expansion of needs is the antithesis of wisdom. It is also the antithesis of freedom and peace. Every increase of needs tends to increase one's dependence on outside forces over which one cannot have control, and therefore increases existential fear. Only by a reduction of needs can one promote a genuine reduction in those tensions which are the ultimate causes of strife and war.

This point has been emphasized again and again in all traditional wisdom. The early Christian mystic Meister Eckhart wrote, 'The more we have, the less we own.' A Sufi religious teacher has been traditionally described as 'he who neither possesses nor is possessed.' Mahatma Gandhi believed that 'the essence of civilization consists not in the multiplication of wants but in their deliberate and voluntary renunciation.'[31]

It is important to understand that the great religious teachings do not promote abject, forced poverty. In fact, the moral and spiritual necessity to redistribute wealth so that all may live decently is proclaimed by all traditional wisdom. What is promoted, however, is restraint, simplicity, voluntary poverty, limits. For if our purpose is to transcend the merely material through contemplation of the divine, possessions and consumption can only clutter our lives by focusing our attention on the transitory, the energy of the world that is constantly degrading. More often than not, the goods we own come to own us. We become attached to them. We fear they will be taken from us. We define ourselves not by who we are but by what we own. In the *Bhagavad-Gita* it is written: 'Thinking of sense-objects, man becomes attached thereto. From attachments longing and from longing anger is born. From anger rises delusion; from delusion, loss of memory is caused. From loss of memory, the discriminative faculty is ruined and from the ruin of discrimination, he perishes.'[32] Or, to put it in painfully familiar modern terms: If you don't own an automobile, you don't need to worry about steel-belted radials, petrol lines, traffic tie-ups, and car thieves.

In a low-entropy culture the individual is expected to live a much more frugal or Spartan life-style. Consumption ceases to be regarded as an end of human existence and reverts back to its original biological function. In the new age, the less production and consumption necessary to maintain a healthy, decent life, the better.

The low-entropy and high-entropy cultures also differ in their approach to labour and production. In a high-energy environment, human labour has no real positive value. The goal of the system is to increase energy flow by eliminating human labour and automating all steps in the production process. Productivity and economic growth become the sole ends of the economy. Where human beings must be involved in the production of goods and services, scientific management is used to remove creativity and individual decision-making by standardizing the method of production. Work, especially physical labour, is considered demeaning, something to avoid. Our society is engrossed with 'labour saving' devices which can remove from human hands all work functions. Pay scales reflect our attitudes toward work: those who labour with their backs and their hands are almost universally at the bottom of the scale; white-collar executives who spend their worktime behind desks are at the top.

In the modern scheme of things, work is a necessary evil, a burden that has to be borne in order to make the money that will allow us to do what we really enjoy. Whenever anyone wins a large sum of money in a lottery, the first thing the press asks is, 'Are you going to quit work now that you're rich?' They are dumbfounded if the recipient of the sweepstakes returns to his job. As for what is produced, that hardly matters at all. The only guiding principle is 'the more the better.' No one takes responsibility for determining whether something should be produced or not. As long as a market for the item can be developed, it will be provided. Thus, society is deluged by a plethora of material effluence – microwave ovens, hair dryers, automobiles that poison the air, and prescription drugs that poison the body.

Whereas industrialism views the end of production as consumption, and work as merely a means to reach that goal, in a low-entropy society work becomes an essential component in our efforts to reach an enlightened state of consciousness. Work in a high-entropy society is secularized; it is divided and measured by the clock and the output; and it is a burden because it has no transcendent significance. In a low-entropy society, human labour is sanctified as any activity that helps us 'know who we really are.' Thus, there is a positive value inhering in work. In his essay 'Buddhist Economics,' E. F. Schumacher suggests that this value is threefold: 'To give a man a chance to utilize and develop his faculties; to enable him to overcome his ego-centredness by joining with other people in a common task; to bring forth the goods and services needed for a becoming existence.'[33]

In low-entropy culture, work is understood to be an activity as necessary for the proper life-balance as sleep, contemplation, or play. Without work, man is incomplete. The person who is engrossed with 'saving labour' and endless indulgence in leisure can no more comprehend the true nature of reality than the person who is lost in the jungle of illusions that comes from an attachment to consumption and possessions.

But not just any kind of work can be considered appropriate. It must be designed, first and foremost, to provide dignity and purpose for the worker. The work must have a human scale to it, a certain type of organization, in order that it can 'give a man a chance to utilize and develop his faculties.' In practice, this means that the type of technology used in work becomes a critical issue in a thermodynamic, as well as a metaphysical, sense. The Entropy Law shows us that the larger the work tools – the machines, the factories – the more capital and energy intensive and the more entropic they become. From the metaphysical point of reference, the scale of the work tools is also critical, for the larger and more centralized they become, the more the role of the human is reduced to just another factor of production. On an automobile assembly line, for instance, the workers must essentially do what the machine 'wants' because the production process is centred on the machine, not the individual. The human loses importance in the work process, and as this occurs, human self-sufficiency is diminished: the worker necessarily becomes dependent on the machines for a livelihood.

Along with the size and type of technology, the organization of production and decision-making also takes on renewed importance. As we have seen, specialization of human activity at the workplace is an outgrowth of a high-entropy economy. The technique of scientific management, the systematic separation of thoughts from action, of conception from performance, is designed to maximize productivity by making non-thinking automatons of workers. Once again, we see that work itself is not valued, only the product of that work.

Likewise, the authoritarian structure of the workplace robs the individual worker of a chance to join in a community with his fellows to make decisions and develop his talents. Unable to join with others to explore his potential and creativity, the individual is forced to retreat into a shell in which he has neither meaningful rights nor responsibilities at his work. All he is left with is a job, a place to make money, and a degrading environment to which he must submit, eight hours of every day.

In terms of governing – both economic and political – a low entropy culture emphasizes the notion that 'the government is best which governs least.' Popular democracy is favoured over rule by the few, and economic arrangements are emphasized in which each person exercises an equal vote and voice in the affairs that affect his life both at the workplace and in the community. Self-managed, worker-run enterprises and small democratically run city-states are the preferred economic and political forms. Decentralized participatory democracy is preferable, not just on a moral or philosophical basis, but also because it minimizes energy flow-through and, as a consequence, reduces the accumulation of disorder. As we've already seen over and over again, highly centralized economic and political institutions only increase the energy flow and the buildup of disorders. Therefore they have no place in an entropic culture based on the limited flow of renewable sources of energy.

In a low-entropy culture the concept of private property is

retained for consumer goods and services but not for land and other renewable and nonrenewable resources. The long-accepted practice of private exploitation of 'natural' property is replaced with the notion of public guardianship. The orthodox economic view that each person's individual self-interest when added up together always serves the common good of the community is regarded with suspicion or, more appropriately, with outright derision. Individual rights are protected, but they are no longer regarded as the dominant reference point from which to judge society. Instead, the notion of public duties and responsibilities once again gains ascendancy as the dominant social motif, as it has been throughout most of history.

In a low-entropy society, our modern view of man and woman divorced from the workings of the ecosystem gives way to a holistic comprehension of the interrelatedness of all phenomena. A low-entropy culture emphasizes man and woman as a part of nature, not apart from it. Nature becomes not a tool for manipulation, but the source of life that must be preserved in its entire workings. Once it is understood that human beings are 'one' with nature, then an ethical base is established by which the appropriateness of all human activity can be judged. For instance, a low-entropic society would view as an obscenity any economic policy that contributed to the destruction of another species. Every species must be preserved simply because it has an inherent and inalienable right to life by virtue of its existence. Because the first law of ecology tells us that 'everything is connected to everything else,' any destruction of one part of nature will affect all other parts, including human beings.

In a low-entropy society the notion of 'conquering' nature is replaced by the idea of harmony with other creatures and the environment as a whole. Like all other forms of life, individual humans are passersby on earth, and have a responsibility to preserve nature to the maximum extent possible, so that those yet to come – both people and other life forms – may enjoy life in their own day.

All of the great teachers of traditional wisdom have embraced the values inherent to a low-entropy life. Buddha,

Jesus, Muhammed, the prophets of Israel, and the mahatmas of India all led exemplary lives of simplicity, voluntary poverty, and communal sharing. Their teachings expressed similar values for all of society. In our own century, Mohandas Gandhi generated an entire liberation movement based on a low-entropy value system.

Above all else, the low-entropy world view shows us the physical limits we face – the limits of our planet's resources and the limits we must impose on the use of technology.

We stand today at the edge of a historic entropy watershed. As we begin the transition from the age of nonrenewable resources to the Solar Age, we will experience much more than a mere shift in the type and amount of energy we use. The movement from a high-entropy to a low-entropy system will transform our values, our culture, our economic and political institutions, and our day-to-day lives. The harbingers of these vast changes are already upon us. Though at this time the signs remain fragmentary, and sometimes contradictory, millions of people are already beginning to conduct their lives in ways that are reflective of the emerging entropy watershed. A 1976 report from the Stanford Research Institute, for instance, estimated that between 4 and 5 million adult Americans have chosen to reduce their incomes drastically and have withdrawn from their former positions as active participants in the high-entropy, industrial, consumer economy. These people have embraced what might most accurately be called 'voluntary simplicity' – a low-entropy form of existence based on frugal consumption; a dominant concern with personal, inner growth over materialism; and a growing ecological awareness. According to SRI, another 8 to 10 million Americans have already adapted parts of this simpler life-style.[34]

Even those more thoroughly wedded to consumerism, industrialism, and urbanism are beginning to make personal adjustments in their lives that seem reflective of the entropy watershed. Whether out of sheer necessity, or choice, these life-style changes are significant steps toward institutionalizing the new world view. We could point to thousands of small signs: a significant growth in urban gardening and neighbourhood food production in cities like Boston, Los Angeles, Washington, DC, and Houston; farmers' markets reopening and prospering in Syracuse, Pittsburgh, New York, and numerous other cities; wood-burning stoves, a few years ago novelty items, now selling out faster than they can be produced; bicycle sales soaring as this transportation vehicle becomes a serious alternative to auto travel; the growth of ecologically minded architectural firms that design homes requiring almost no heat or air conditioning; alternative-technology corporations springing up around the country; a burgeoning cottage handicraft industry whose wares are daily sold on street corners in Seattle, Chicago, and Philadelphia. All of these, of course, are but fragments of the new world order, but they point the way.

In a more organized fashion, the nascent anti-nuclear-power, pro-solar-energy movement and the largely unstructured, but growing, human potential and New Age politics community in America are already developing a synthesis of political programmes and personal life-style choices that anticipate the entropy watershed. Both groups, for instance, have already largely rejected the mechanistic world view. Author-activist Mark Satin speaks for many in both the solar movement and the New Age constituency when he writes of the modern scientific outlook:

It has encouraged us to lose touch with ourselves and our bodies; it has cut us off from other dimensions of reality besides the material; it has led to our worship of machines and of technique; it has led to a situation where the human scale is lost, and 'progress' means mostly destruction; it has helped us forget that after all the 'objective' facts are in, we still have to make moral choices and value judgments; it has led to a separation of means and ends in almost every human endeavour.[35]

But, having rejected the prevailing machine world view, neither group has yet developed an all-embracing philosophical alternative. To date, only bits and pieces of their alternative vision have been formed. In the coming years, the Entropy Law will be increasingly embraced by solar and New Age activists as the context for their alternative visions.

In many ways, these two activist movements are a reflection

231

of the dramatic shift that has taken place in American public opinion in the past ten years. Dissatisfied with the unworkability of highly centralized economic and political institutions, and suspicious of technological developments that promise much but always seem to end in social, economic, and environmental disruptions, Americans are voicing opinions that are increasingly antithetical to a high-entropy world view. Among the findings of a Harris poll in May 1977, for example, were the following: by a margin of 79 per cent to 17 per cent, the public would place greater emphasis on teaching people how to live more with basic essentials 'than on reaching a higher standard of living'; by 76 per cent to 17 per cent, a sizeable majority opts for 'learning to get our pleasures out of nonmaterial experiences' rather than 'satisfying our needs for more goods and services'; 59 per cent believe we should be 'putting a real effort into avoiding those things that cause pollution' over 'finding ways to clean up the environment as the economy expands'; 82 per cent would prefer to 'improve those modes of travel we already have,' while only 11 per cent believe we should 'develop ways to get more places faster'; by 77 per cent to 17 per cent, Americans would prefer 'spending more time getting to know each other better as human beings on a person-to-person basis' instead of 'improving and speeding up our ability to communicate with each other through better technology.' Finally, nearly two-thirds of the public say that 'finding more inner and personal rewards from the work people do' is more important than 'increasing the productivity of the work force'; 'breaking up big things and getting back to more humanized living' should take precedence over 'developing bigger and more efficient ways of doing things'; and 'learning to appreciate human values more than material values' should be placed above 'finding ways to create more jobs for producing more goods.'[36]

These startling attitudes reflect the schizophrenic state of Americans today. Trapped in their roles as active, daily participants in the high-energy culture, the vast majority of people indicate that what they really value are the ideals that are most compatible with a low-entropy environment. Imagine what would happen to our social and economic system if these beliefs were taken out of the realm of personally held values and fully integrated into our society! The entire system would be transformed at its very base, and the Newtonian machine paradigm would be a thing of the past.

It would be misleading, however, to read too much into these hopeful signs. The transition to the Solar Age will not be accomplished easily. Because our society has been designed to maximize energy flow, the dawning of the new energy environment will require a monumental disruption in the current order of things. Sacrifice and hard work will be necessary to make it through the transition period. Writing of another empire at another time, C. W. Hollister, professor of medieval history at the University of California at Santa Barbara, has perhaps outlined elements of our own fate. The fall of Rome, he notes, 'brought disorganization and savagery, but it also gave Europe the chance for a new beginning, an escape from old customs and lifeless conventions, a release from the stifling prison that the Roman Empire, for most of its inhabitants, had come to be . . . Life in the post-Roman West was hazardous, ignorant, foul and deeply insecure, but such was the price of the new beginning. Periods of momentous change are seldom comfortable.'[37]

As we look ahead to the emerging order, we are naturally eager to see how our society will be arranged and how our lives will be changed. We cannot, of course, divine all of the specifics, any more than Locke, Bacon, and Adam Smith could have forecast how their philosophies would come to fruition 300 years later in the modern technological society of twentieth-century America. Nonetheless, using the general principles of the low-entropy paradigm, we can already foresee the bare outline of the tremendous social reformation that awaits us.

In the Solar Age, agriculture will be transformed into diversified organic farming. Organic farming uses no chemical fertilizers and pesticides, but relies instead on natural manures and natural pest enemies. Studies done comparing organic farming with chemical farming show that while the yield per acre is roughly the same, organic farming uses two-thirds less energy. Organic farms use 6,800 BTUs of energy to produce a dollar of output, whereas conventional farms use over 18,400 BTUs. One study found that while the cost of conventional farming – using highly mechanized farm machinery and massive doses of chemical fertilizer and pesticides – averaged $47 per acre, the cost per acre on an organic farm was only $31.[38] As the cost of energy skyrockets in the years ahead, organic farming will prove to be an even more economically viable alternative; not to mention the fact that organic farming produces crops of higher nutritional value and creates less pollution in the environment.

Large-scale centralized farming will also have to give way to the kind of small-scale regional truck farming the United States had before World War II. The energy costs in transporting agricultural products across the country to distant markets will soon be so high that local and regional farming will provide a less costly economic alternative. Farmers' markets, which once existed in towns and cities nationwide, are already making a dramatic comeback. Squeezed by the escalating middleman costs in food processing, farmers and consumers are beginning to deal with each other directly.

Small-scale labour-intensive agriculture will require a shift of people away from the cities and back to the farms. The transition will not take place overnight, but rather gradually over an extended period of time. Eventually the proportion of farm to city population will have to reverse itself if human life is to survive. Labour-intensive organic farming simply cannot support the concentrated urban population centres that have built up during the high-energy fossil-fuel age. An agricultural way of life will dominate the coming Solar Age as it has in every other period of history before our own.

While urban life will not disappear, the reign of the megalopolis will come to an end. 'Large' cities will once again return to their preindustrial size of 50,000 to 100,000 citizens. Not only will this be in keeping with the ability of the surrounding environment to produce both food and solar power, but numerous studies in recent years have indicated that when urban centres grow to more than 100,000 residents, disorders also grow at an alarming rate. As we saw earlier, large cities are disproportionately costly to run and have a much higher incidence of crime, mental illness, pollution, and other forms of disorder. In the Solar Age, big cities will simply be too large a drain on precious resources.

Along with the scaling down of cities, transportation systems are also going to be vastly reoriented in the years to come. The high cost of energy is going to force a fundamental shift in the pattern of travel away from automobiles and trucks and toward greater mass transit and long-distance rail use. Bicycling and walking are also going to become increasingly popular modes of travel. Our social and economic life will undergo radical changes reflecting the change in transportation. Less recreational time will be spent in travel, and more near and around home. Places of business will increasingly be located within a short travel radius of available labour pools. There will also be a move away from large centralized institutional complexes serving widely spread out population areas. The neighbourhood community centre will replace the giant shopping centres and other institutions that have sprouted up alongside the auto age.

Industrial production and the service sector will be greatly reduced in scale in the low-entropy economy of the future. This will be due to the necessity of transferring much of the labour force to food production, as well as to the impossibility of maintaining the high energy flow required for the maintenance of the industrial-service infrastructure.

Our modern economy is a three-tiered system, with agriculture as the base, the industrial sector superimposed on top of it, and the service sector, in turn, perched on top of the industrial sector. Each sector owes its existence to the increased flow of nonrenewable energy. The increased energy flow made possible the replacement of labour-intensive with capital-intensive processes in the agricultural sector, creating tremendous economic surpluses and freeing vast numbers of workers to pursue alternative employment. The surpluses and the excess labour pool provided the base for the emerging industrial sector. Likewise, the increased use of nonrenewable energy in the industrial sector made possible a similar shift from labour to capital-intensive processes, creating additional economic surpluses and freeing millions of workers for employment in a new arena, the service sector.

Given the 'placement' of the three major economic sectors of society, it is not hard to predict the likely future course that our economic system will follow. As energy slows its flow through the economy, funds will be diverted away from both the public and private service sectors. Employment in service areas will be the first to suffer, because services are the least essential aspect of our survival. With the contraction of the service sector, unemployed workers will seek out jobs in industry. Because of escalating energy and resource costs, industry will reverse its historical trend and convert back from energy- and capital-intensive production modes to labour-intensive ones, thus absorbing some of the surplus labour force from the contracting service sector. (At the same time, agriculture, which will no longer be able to continue its mechanized farming techniques, will also become far more labour intensive, absorbing more of the surplus work force cast adrift from the service sector.)

In keeping with the dictum that the low-entropy economy is one of necessities, not luxuries or trivialities, production will centre on goods required to maintain life. To recognize the extent to which production will be diminished, we have only to take a tour through a suburban mall and ask ourselves, 'How many of these products are even marginally useful in sustaining life?' Any honest appraisal is sure to conclude that most of what is manufactured in our economy is simply superfluous.

The production that does continue should take place within certain guidelines in keeping with the low-entropy paradigm. First, production should be decentralized and localized. Second, firms should be democratically organized as worker-managed companies. Third, production should minimize the use of nonrenewable resources. All of these points are consistent with both the energy and ethical requirements of the entropic world view. Of course, adhering to these guidelines will necessarily mean that certain items will become impossible to produce. A Boeing 747, for instance, simply cannot be manufactured by a small company employing several hundred individuals. Thus, a new ethic will have to be adopted as the litmus test of what should be produced in the low-entropy society: if it cannot be made locally by the community, using readily available resources and technology, then it is most likely unnecessary that it be produced at all.

Many industries will not be able to withstand the transition to a low energy flow. Unable to adapt to the new economic environment, the automotive, aerospace, petrochemical, and other industries will slide into extinction. Many of the workers will need to retrain for new labour-intensive trades vital to the survival of local communities. But again, we should not be beguiled into thinking that the transfer of workers from one mode of industrial production to another can be made easily. Like it or not, the shift in the mode of economic organization will mean hardship and sacrifice.

The move toward a low-entropy economy will spell the end of the reign of the multinational corporation. There are many reasons that these corporate behemoths will not be able to withstand the changing energy environment. They are too complex, and they are completely reliant for their maintenance on the extracting of nonrenewable resources from all over the world. The multinational corporation is the dinosaur of our energy environment. Too big, too energy consumptive, and too specialized, they will run into their own evolutionary dead end as production moves back to a localized, small-scale base.

The uses of technology will also change drastically in the future. Once technology is recognized as being essentially a transformer of energy from a usable to an unusable state, we will come to understand that the less we use complex energy-consuming technologies, the better off we are.

In a low-entropy society, big, centralized, energy- and capital-intensive techniques will be discarded in favour of what is called appropriate or intermediate technology. Futurist-author Sam Love defines appropriate technology as 'locally produced, labour-intensive to operate, decentralizing, repairable, fuelled by renewable energy, ecologically sound, and community-building.'[39] E. F. Schumacher, credited as the father of the intermediate-technology movement, says that this low-entropy form of technique is 'vastly superior to the primitive technology of bygone ages but at the same time much simpler, cheaper, and freer than the super-technology of the rich. One can also call it self-help technology, or democratic or people's technology – a technology to which everyone can gain admittance and which is not reserved to those rich and powerful.'[40]

Finally, the low-entropy age we are moving into will require a great reduction in world population. The massive explosion in world population is really only understandable when viewed in thermodynamic terms. Picture our world at its beginning, before life began its development and evolution. The planet was covered with seas, mountains, and valleys. Then, 3 billion years ago, life began – growing literally out of the energy resources contained within the earth and emanating from the sun. Like all other forms of life, when Homo sapiens appeared on the planet some 3 million years ago, he sustained himself by receiving energy from the renewable resources of the sun. Because life was based on this decentralized flow of solar power, the absolute numbers which could be sustained remained relatively low. Population density increased quite slowly. It took the entire period of human existence until roughly 1800 to reach the first billion people.

From that point on, the explosion began in earnest. As we saw earlier, the second billion humans took only a hundred years. The third billion took only thirty years, between 1930 and 1960. The fourth billion took only fifteen years. At current growth rates, the world's population will double once again to 8 billion by the year 2015 and to 16 billion by the year 2055. This explosion in population corresponds exactly to the world's shift from an agricultural economy (based on solar flow) to an industrialized system (based on the extraction of a nonrenewable stock of energy from the earth's capital). In other words, it is not only our buildings, cars, and other artifacts that are made from fossil fuels and nonrenewable resources: in a sense, the 4·5 billion additional people on earth today have also resulted from this massive transformation of stored energy; that is, the massive population surge has

238

been made possible by the conversion of billions of years of stored solar energy. It should come as no surprise that the industrial era based on nonrenewable resources comprises less than ·02 per cent of human history, and yet '80 per cent of the increase in human numbers has occurred during this period.'[41]

The implications of a thermodynamic view of population growth are staggering. In the preindustrial solar age, the carrying capacity of the world, in terms of human beings, was only 1 billion. Even at that, the world's resources were being severely strained. As a direct result of a higher energy flow of stored nonrenewable resources, 3·5 billion additional people were added to the world's load. Without this energy flow, these people could not be sustained. Yet, as we have seen, the finite limit of our planet's resources makes it impossible that the energy flow of the past 200 years can long continue. Because of this, it is essential that the world begin with renewed vigour a serious programme aimed at reducing the earth's population in the decades to come. The world must once again move back toward a sustainable, Solar Age population.

The world's population will diminish, of that there can be no doubt. The question is how. There have been many specific proposals: licensing parents to have no more than two children; changing tax laws so that people who have children are penalized severely for each additional offspring; programmes of force such as that in which 11 million Indians were sterilized under Indira Gandhi's rule. All of these programmes are distasteful, at best, because they are externally applied by society. The only other alternative is a full internalization of the entropy paradigm, so that we voluntarily limit our population by exercising restraint in our individual desires to have children. Once we fully comprehend that each child we bring into the world places a burden on succeeding generations by denying them their own share of resources to sustain their own lives, then we can develop a set of values that will lead to a humane programme of population control.

For most Americans, any discussion of population control immediately conjures up Malthusian visions of India, China, and other overcrowded Third World nations. It is, of course, urgent that these poorer countries move to substantially curtail their own populations. But lest we think that the population problem is endemic to the Third World alone, let us remember that it's not just the numbers of people who physically exist that are important, but also the amount of energy consumed by each individual. As we've seen before, in the United States we are using as much of the earth's fixed energy resources as 22 *billion* people. A population map of the planet based not just on human numbers but on energy consumption would show that the greatest population problem in the world today, in terms of energy depletion, exists right here in the United States. Thus, we must not only move to limit the absolute numbers of individuals in our country, but we must also limit our energy consumption drastically.

Our list contrasting high-entropy and low-entropy social systems could go on and on. In even this brief overview of the general nature of the emerging entropic society, it should be clear that vast transformations are impending. From our perspective, the coming changes may not appear desirable. Let's face it: most of us, having lived in an era of unparalleled material affluence, and indoctrinated by education, television, and advertising, are hedonists to one degree or another. Nicholas Georgescu-Roegen wonders whether we will be able to wrest ourselves from our existing world view:

Will mankind listen to any programme that implies a constriction of its addiction to exosomatic comfort? Perhaps the destiny of man is to have a short, but fiery, exciting and extravagant life rather than a long, uneventful and vegetative existence. Let other species – the amoebas, for example – which have no spiritual ambitions, inherit an earth still bathed in plenty of sunshine.[42]

If the task ahead seems impossible to accomplish it's only because we continue to view what needs to be done with Newtonian eyes. Our existing world view cannot possibly provide us with the confidence and zeal we need to overcome the present historical crises, because it is bound up with the existing energy environment. Only the entropy paradigm provides a scythe that is both sharp enough to cut through the

240

tangled debris of this death-bound culture and broad enough to clear a path for the dawn of a new age.

The specifics of what is to be done can only come after a thorough cleansing away of the worst remaining vestiges of the mechanical world view. Our own conversion is the first order of business. Only when we have cast aside forever the old way of thinking and behaving and take onto ourselves the new entropic world view will we be ready to go forth and remake our culture. The new order of the ages must begin with a revolution in science, education, and religion. In each area, the old mechanical constructs must be replaced by new constructs cast to the requirements of the second law.

Reformulating Science

It is a curious fact that, just when the man in the street has begun to believe thoroughly in science, the man in the laboratory has begun to lose faith. When I was young, most physicists entertained not the slightest doubt that the laws of physics give us real information about the motions of bodies, and consist of the sorts of entities that appear in the physicists' equations.[43]

These words come from Bertrand Russell's pen. If the people knew what the physicists now know, the bottom would fall out from the mechanical world paradigm. The assumptions of classical physics upon which we have confidently erected our entire way of organizing life turn out to be largely fallacious, say today's scientists.

Take, for example, the idea first expounded by Descartes that the world could be understood and then organized by the 'scientific method,' that is, the separation of things into subjects and objects that could be precisely measured and quantified by mathematical formulas. Quantum theory says it just isn't so. Early in the twentieth century, scientists began peering farther and farther into the microworld of life, trying to locate, isolate, and measure the most elemental particle of matter in the universe. They found that as they probed deeper and deeper, locating more minute elements along the way, that there seemed to be no end in sight. Then it dawned on them that the whole procedure was like a giant cosmic joke unfolding at their expense. The scientific community was red-faced, to say the least, when the German physicist Heisenberg discovered that the 'objective observation of atomic particles was an impossibility, the very nature of atomic particles being such that the very act of observation was interfering and altering, instead of fixing and preserving, the object.'[44] Heisenberg and those who followed him into the microworld of quantum physics learned, with each new observation they made, that precise measurement of matter –

the very basis of classical physics – is impossible because it calls for being able to determine both the velocity and location of an object at a given moment in time. To their chagrin, they realized that every time they observed the tiniest of particles, the electron, their act of observation was influencing what they saw. That's because 'you can only see an electron when it emits light, and it only emits light when it jumps, so that to see where it was "you" have to "*make it*" go elsewhere.'[45] That being the case, you can't have it both ways. You can measure either its location or its velocity but never both at the same time. The point is, 'If you know where you are, you can not tell how fast you are moving, and if you know how fast you're moving, you cannot tell where you are.'[46]

Heisenberg's discovery was given a name: the Heisenberg uncertainty principle. His discovery marked the darkest day in the history of classical physics. For all intents and purposes he pulled the rug out from under the ironclad determinism that had surrounded the laws of physics for nearly 300 years. In science, only one uncompromising exception is enough to invalidate a law. Heisenberg broke the back of Newtonian science and the world view that was built upon it.

Heisenberg's uncertainty principle, however, was only the opening salvo of a long and protracted scientific assault that has left much of classical physics in shambles. Newton's claim that he had found the scientific key for unlocking the secrets of the universe is now treated as little more than the quaint bravado of an infant science, unencumbered by the paradoxes and complexities that ultimately accompany the process of growth in knowledge.

Up to a hundred years ago, physics rolled along self-assured by its own claims that any set of initial conditions can lead to one, and only one, final state. Today, the causality principle of classical physics has become so qualified that it could hardly claim the status of a law. Scientists now acknowledge that a given set of initial conditions can lead to several possible alternative states. They distinguish between the early deterministic laws and the newer, indeterministic laws. In the latter case, probabilities are assigned to each of the possible outcomes of an initial set of conditions as the best

243

that can be hoped for in terms of measurement. But even the indeterministic laws are now being challenged by what some scientists refer to as 'a second stage of indeterminism,' in which the assignment of probabilities for the various outcomes of an event are virtually impossible to establish. The renowned physicist Max Born summed up the frustration of his colleagues over the direction in which their own research has led: 'We have sought for firm ground and found none. The deeper we penetrate, the more restless becomes the universe; all is rushing about and vibrating in a wild dance.'[47]

What the scientists have learned is that every event is unique; its own occurrence distinguishes it from all other events. For that reason, each event not only claims a place all its own in the world but cannot be said to share an objective reality with any other phenomenon. Its subjective occurrence, in turn, is not the result of a particular initial set of conditions. Rather, it owes its occurrence to the entire labyrinth of all past subjective occurrences whose collective configuration gave rise to its own particular unfolding. The idea that specific phenomena can be isolated from the rest of the universe they're a part of and then connected in some kind of 'pure' causal relationship with other isolated phenomena is just plain wrong thinking. The Newtonian paradigm of precise measurement, of dividing matter into neat quantities that can then be linked to each other and rearranged without regard to either their effect on the rest of the cosmos or the effect of the rest of the cosmos on them, has led to the wanton manipulation and destruction of nature at the hands of modern science.

Everything in this world is connected with everything else in a delicate and complex web of interrelationships. The best computer ever designed by humankind still cannot calculate even a tiny fraction of all the relationships that exist in the ecosystem of a simple pond. Scientists have tried it and have only thrown up their hands in despair after realizing the complexity and detail involved.

The old Newtonian view that treats all phenomena as isolated components of matter, or fixed stocks, has given way to the idea that everything is part of a dynamic flow. Classical physics, which recognized only two kinds of classifications, things that exist and things that don't exist, has been challenged and overthrown. Things don't just 'exist' as some kind of isolated fixed stock. This static view of the world has been replaced by the view that everything in the world is always in the process of becoming. Even nonliving phenomena are continually changing. This process of becoming is really nothing more than the Entropy Law at work. Every single thing is energy and that energy is continually being transformed. Every transformation affects everything else that is in the process of becoming. The life and death of every blade of grass affects the total change in the energy in the world. As mentioned earlier, the Entropy Law tells us the direction in which the energy flow moves, but not the speed. The speed fluctuates. There is nothing smooth about the ebb and flow of the becoming process. It moves along in jumps and spurts.

How different this scientific view is from Newtonian physics with its simple matter in motion, its fixed forces acting against other fixed forces in precise and predictable ways. It's no accident that a science based on manipulating fixed stocks is being replaced by a science based on understanding dynamic flows, now that we are moving from an energy environment based on stocks (fossil fuels) to one based on flows (the sun's rays, and renewable resources). The scientific assumptions are changing to reflect the realities of the new energy environment.

Ilya Prigogine, who received a Nobel Prize for his work on nonequilibrium thermodynamics in 1977, says that the entire notion of causality and precise measurement, the hallmarks of classical physics, are about to give way to a redefinition of science based on the imperative of the second law. Every occurrence in the world is unique, argues Prigogine, and for that reason it's impossible to make precise predictions about the future based on scientific observations. The most that science can do is predict likely scenarios. The old security provided by classical physics was an illusion from the beginning, say Prigogine and his colleagues. It is not possible to know nature in the sense that Descartes, Bacon, and Newton had in mind. The idea that human beings can separate themselves from nature, discover its inner secrets, and then use them as a 'fixed body of truths' to manipulate and change the natural world has proven to be erroneous. First, as scientist Niels Bohr once remarked, we are all actors as well as spectators in the unfolding of the natural order. We can't separate ourselves from the world around us no matter how hard we try. Secondly, the notion of fixed bodies of truths, in the deterministic sense of classical physics, no longer holds up as we now experience a universe of continual fluctuation and instability. Prigogine captures the essence of the new reformulation of science when he says that 'instead of the classical description of the world as an automaton we go back more to the Greek paradigm of the world as a work of art.'[48]

In the final analysis, every science is nothing more than a methodology for predicting the future. At the same time, every scientific methodology is involved in a constant search to define the upper limits of what is possible. A scientific law remains valid as long as it satisfactorily predicts the future and no exception can be found that undermines the limits it establishes. The Entropy Law satisfies both objectives. More than any other concept yet discovered, the Entropy Law provides a comprehensive methodology for predicting the future and establishes the supreme limits within which things can take place in this world.

The Entropy Law will soon supersede Newtonian mechanics as the ruling paradigm of science because it, and only it, adequately explains the nature of change, its direction, and the interconnectedness of all things within that change process. The Entropy Law may someday be invalidated and overthrown. But for now it remains the one law of science that seems to make common sense out of the world we live in and provides an explanation of how to survive within it.

Reformulating Education

Our entire learning process is little more than a twelve-to-sixteen-year training programme for the Newtonian world view. In school, emphasis is placed on quantities, distance, and location but rarely on qualities or conceptions. Think of all of the countless tests where the only questions asked were those concerning names, dates, and places – things that could be precisely measured and that involved no ambiguities. The tests themselves are cut directly from the mould of classical physics. True, false, fill in the blanks, multiple choice, and matching answers are all based on the concept of causality; that for every set of initial conditions there is one and only one correct final state. The most important aspect of taking tests is not the answers but the process. We all forget specific facts over the years, but few of us will ever forget the concept of causality after being subjected to the testing process for so many years of our life.

If a poll were taken, it's likely that just about everyone who's gone through school will recall a time when he or she questioned the testing procedure itself. How many of us have had the experience of taking a test and looking at the specific question, and all of the specific answers, and feeling, some-how, compromised by having to pick one of them. Our common sense told us that it wasn't as simple as all that. We told ourselves that other things needed to be considered, and that it was just plain foolish to try to isolate this particular phenomenon from everything else around it. Still, after a bit of quiet mumbling, we allowed ourselves to succumb to the process. If we had to pick one answer, then so be it. We might even rationalize our surrender by musing that even if there was no right answer, at least we could select the one best answer.

At this very moment children all over America are taking tests or preparing for them. What they don't realize is that what they're really learning is not just facts but how to think in terms of causality and quantification, the basics of the Newtonian world paradigm. When our educators claim they are teaching children how to think, this is the particular type of thinking they have in mind. Of course, few of them are conscious of the 'fact' that they are promulgating a particular ideology when they teach. They would probably protest that their only concern is to teach the child how to think 'objectively.' Need we say more?

The thinking process is only important if it produces results, and that means learning facts. Our educational system places its highest priority on facts. The more bits of information a student can collect and recall, the better grades he will receive. Facts are valuable, it is argued, because they help one to better understand the world and to better organize one's own life. The amount of *facts* we know about the world around us is doubling every few years. Yet one would be hard pressed to claim that the world is becoming twice as well organized as a result. As we know, the opposite is the case.

It has been said, 'Show me a person who has command over the facts and I'll show you a person in control of the situation.' If that sounds suspiciously close to what Descartes, Newton, and Bacon might have said, it's only because it's a 'fact.' It's also true that we continue to rummage through the world collecting more and more facts, despite the evidence that it only adds to the confusion and disorder piling up around us. The reason we do it is that we just can't free ourselves from the influence of those gentlemen of yesteryear and their outworn ideas.

Finally, our educational process is devoted to specialization. Every time we learn something new and different about the world, a new academic or professional discipline is set up to collect and interpret the new data. Learning has become fragmented into tinier and tinier frameworks of study on the Newtonian assumption that the more we know about the individual parts, the more we will be able to make deductions about the whole those parts make up.

Visit any major university complex and you will see thousands of men and women walking to and from laboratories and classrooms each with his or her own special briefcase – with labels ranging from political science to sociobiology. Each of those briefcases will be crammed with lots of facts and figures about a specific phenomenon that have been rearranged to fit the criteria of the particular discipline in which they are working. From the specific data and within the constraints imposed by their specific discipline, they confidently set forth their views about the way part or all of the world works. The cardinal sin among academics is fraternization with the enemy; any scholar worth his credentials would never even consider cross-checking his notes with those of other disciplines. An interdisciplinary or general approach to learning is labelled 'not serious.'

Our professionals have become like thousands of little blind creatures poking their sticks furiously at different parts of the elephant, each with a different notion of what the beast must look like. The more they poke at the little space reserved for them, the more convinced they are that they know what they are poking at, and the more wrong they become.

Our educational process is designed to accommodate the needs of an industrial society. Industrial society, in turn, is designed to suit the needs of a nonrenewable energy base. As we begin to make the transition to a solar energy environment, our current approach to education and learning will be rendered increasingly obsolete. The Newtonian style of learning will be forced to give way to an entropic approach to education.

The emphasis in learning will be dramatically opposed to the way we go about things today. For example, education will stress process over measurement. The notion of collecting, storing, and exploiting *stocks* of isolated facts will be replaced with the idea of examining the *flow* of interconnected phenomena. Testing will focus on conceptual abilities over empirical ones; and essays, oral discourse, and practical experience will be the standard forms reflecting the need to think in terms of process. The external world will be examined not as a series of isolated causal relationships, but as a web of interrelated phenomena expressing many possible scenarios for movement and change.

Collecting facts will be less important in the Solar Age because the emphasis will shift from exploiting nature to living with it. Education, like science, will be more concerned with the *why* of things as opposed to the *how*. The shift from the empirical to the metaphysical will mean a corresponding reduction in the information or energy flow-through and in the subsequent disorders that result from it. Learning, then, will not be seen as a tool to carve up the world and fashion it into something else, but as a method to better understand how to live within the limits of the world we've inherited from nature and which we are a part of. Learning as progress will be replaced with learning as the process of becoming.

While some specialization will be required, even in the Solar Age, the educational process will be centred on a holistic approach to knowledge. Unlike today's educational system, which separates students into either trade or academic tracks in high school and afterward, the new emphasis will be on combining mental and manual skills, teaching each person to become self-sufficient in the world. This approach to academic training will be absolutely essential, as work roles in society will be far less specialized in the world that is emerging. Students will be prepared for more general labour-intensive jobs in society and will be expected to be jacks-of-all-trades in order to help maintain smaller, self-sufficient urban and rural communities.

The artificial separation between human culture and nature, characteristic of the Newtonian era, will give way to a new reunification of the two in the coming Solar Age. The concept of 'man against nature' will be replaced by the concept of 'people in nature.' The educational process will reflect this basic change. In contrast to the current academic process, which separates students from the outside world for twelve to sixteen years in a hermetically sealed, artificial environment, the educational experience in the entropic era will emphasize learning through day-to-day experiences in the world. Apprenticeship will once again take on the importance it has had in previous periods of history. At the same time, the large, centralized learning complexes typical of the last stages of the age of nonrenewables will give way to the notion of 'learning environments.' In the Solar Age, going to school will mean going into the community to learn.

Much of the knowledge accumulated during the Industrial Age will become increasingly irrelevant in the coming Solar Age, and will eventually be discarded altogether. However, some bits and pieces of information will remain useful and will be retained and passed on through the educational process. Every major shift in world view incorporates fragments of the old schematic within the new order. While many of the distinguishing features of the old world view will survive as a part of the new paradigm, their role and importance will be radically redefined to fit the new set of governing assumptions.

Although the emerging educational process is likely to twist and turn in many new and as yet unimaginable directions, it will be guided, from beginning to end, by the overriding principles of the first and second laws of thermodynamics.

A Second Christian Reformation

The emerging entropic world view is already being accompanied by a radical reformulation of Christian theology. The Protestant Reformation, which provided an expansionary theology ideally suited to the expansionary economic era of the past 400 years, is giving way to a new theological construct, one reflecting the requirements of the Entropy Law and the new Solar Age.

Over the past fifteen years there has been a great deal of experimentation in this country with Eastern religions. Today, over half a million Americans are adherents of Buddhist theology, and 4 to 5 million more practise meditation, yoga, and other mental and physical exercises whose source of inspiration is found deep in the religious experience of the East.[49] At the same time, the United States is experiencing a massive evangelical revival, which pollsters like George Gallup are claiming represents the early stage of a third great awakening in America.[50]

America has experienced two other great religious awakenings in the past. The first great spiritual revival, in the 1740s, helped unite the colonies and served as a catalyst in the political movement against the crown. The second such awakening, a century later, helped spawn the abolitionist movement and set the stage for the Civil War. Today, the evangelical fervour is spreading across the land once again, and there is every reason to believe that this third great awakening will help spark a profound change in the social and economic life of the nation, just like the first two revivals.

The growing interest in Eastern religions and the mushrooming evangelical movement represent the unconscious search for a new religious synthesis that can accommodate the new age we are moving into. Each brings with it an essential ingredient for a new theological reformulation.

Adherents of the Eastern religions – and especially the

Buddhists – have long understood the value of minimizing energy flow-through. The practice of meditation is designed to slow down the wasteful expenditure of energy. The state of Nirvana or truth is reached when the individual is expending the least energy necessary to support his outward physical survival. The Eastern religions have long claimed that unnecessary dissipation of personal energy only adds to the confusion and disorder of the world. Ultimate truth, according to Eastern doctrine, is arrived at only by becoming one with the world around you. This can only be accomplished by entering into a unified relationship with the rest of nature.

Westerners have always had a hard time understanding the Eastern approach to truth and wisdom. We have believed that only by constantly doing can we unlock all the hidden secrets of the world. So we are perpetually engaged in collecting and piecing together bits and fragments of truth and manipulating and rearranging the world around us, convinced that our efforts will lead to increased wisdom and eventually bring us face to face with the supreme architect of the universe. The Eastern theologians would say that our frenetic activity is only increasing the disorder and confusion and removing us further away from the divine revelation we seek.

While the Eastern religions have understood the value of minimizing energy flow and lessening the accumulation of disorder, it is the Western religions that have understood the linear nature of history, which is the other important factor in synthesizing a new religious doctrine in line with the requirements of the Entropy Law. Unlike Eastern theology, which emphasizes recurring worlds and history as cycles, the Judao-Christian tradition has always taught that earthly history has a distinct beginning and end.

On the other hand, the traditional Christian approach to nature has been a major contributing factor to ecological destruction.[51] The overemphasis on otherworldliness has led to disregard and even exploitation of the physical world.[52] This view holds that the only things of true value are those found in the heavenly world of God. Our world, the world of people and nature and the flesh, is seen as low, depraved, and unworthy and therefore of little concern or consequence to those seeking to live a holy life. The natural world is merely a stopover on our journey to the next world. Therefore, the less attention placed on it and the more attention placed on God's kingdom, the better.

The other shortcoming of Christian doctrine over the centuries has been the interpretation of the concept of dominion in the account in Genesis of the Creation: 'Be fruitful and multiply and fill the earth and subdue it; and have dominion over the fish of the sea and over the birds of the air and over every living thing that moves upon the earth.' The concept of *dominion* has been used by people to justify the ruthless manipulation and exploitation of nature. Now, however, a major reformulation of Christian doctrine is beginning to take shape. For the first time, Christian scholars are beginning to redefine the meaning of dominion, and in so doing they are creating the theological foundations for an entropic world view.[53]

The new interpretation of Genesis begins with the idea that since God created the heavens and the earth and everything in this world, all his creations take on importance and an intrinsic worth because they are of his making. Since this creation of God's has a purpose and order to it, that purpose and order is to be revered just as God's creations are to be revered. Finally, what God has created is fixed. The Lord created the world and *everything* in it and then he rested, according to the Creation story. It follows from this, argue the new theologians, that anything that exploits or harms God's creations is sinful and an act of rebellion against God himself. Likewise, anything that undermines the fixed purpose and order that God has given to the natural world is also sinful and an act of rebellion. This is no small theological point. Every other religious conviction flows from these central truths of Creation, contend the new theologians. Either God created the world or he didn't. Either God gave purpose and order to the world or he didn't. If one believes in these truths, then one believes in God. If one doesn't believe in these truths, one can't possibly believe in God. This thesis is the beginning point for all Christian believers.

It follows, then, that sin is people's hubris in believing that they can treat God's creations differently than God does; namely, manipulate and exploit them for purposes other than what they were created for. Sin is also people's hubris in believing that they can reorder this world and redefine its purpose to suit their own whims and fancies. The Christian life must be one of conserving wholeness over fragmentation, balance over imbalance, and harmony over disharmony. A Christian must love God's creation and treat it with respect because God created it with love.[54]

Dominion, then, does not mean the right to exploit nature. Far from it, say the scholars. Dominion means stewardship over nature. Henlee H. Barnett, in his book *The Church and the Ecological Crisis,* points out that the Biblical view of humankind 'is that of a keeper, caretaker, custodian . . . of the household earth.' Stewardship, says Barnett, is 'the New Testament term for this role of human beings in relation to the natural order.' The first requisite of a steward, according to Barnett, 'is faithfulness, because he handles that which belongs to another.'[55] The concept of stewardship leads directly to the Biblical notion of covenant. In Genesis, God says, 'I established my Covenant with you [humankind], and with your seed after you and with every living created thing.'

God, then, has a covenant with humanity. Men and women are to act as His stewards on earth, preserving and protecting all of God's creations. This covenant puts human beings in a special relationship to God. Since people are a creation of God, just like all of God's other creations, they are equal to them in their finite nature; only God is infinite. While all creations are equal in that they owe their existence to the same source – God – human beings are nonetheless different. The difference, as Francis Shaeffer points out in his book *Pollution and the Death of Man,* is that human beings are made by God in his image and are given the responsibility to act as stewards over the rest of God's creation. Therefore, people are both part of nature, equal to and dependent on all other living and nonliving things, and at the same time separate from nature with a responsibility to protect and take care of it. As long as people accept both relationships, they are faithful to God's purpose and are carrying out the covenant God

255

made with them. However, when people take advantage of their special relationship by taking over God's creation as their own, using it for their own ends rather than God's glory, they have broken the covenant and are rebelling against God.[56]

The new stewardship doctrine and the laws of thermodynamics, when combined with more orthodox theology, set the tone for a new, reformulated Christian doctrine and covenant suited to the ecological prerequisites of an entropic world view. Most of all, the stewardship doctrine provides an answer to the ultimate question, 'Why should I take the responsibility of caring for and preserving the natural order?' Because it is God's order. God created it and God entrusted human beings with the responsibility of overseeing it. It comes down to a question of serving God or rejecting Him.

The new stewardship doctrine turns the modern world view upside down. The rules and relationships that are used to exploit nature are diametrically opposed to those that are necessary to conserve nature. For example, private ownership of resources, increased centralization of power, the elimination of diversity, greater reliance on science and technology, the refusal to set limits on production and consumption, the fragmentation of human labour into separate and autonomous spheres of operation, the reductionist approach to understanding life and the interrelationships between phenomena, and the concept of progress as a process of continually transforming the natural world into a more valuable and more ordered human-made environment have long been considered as valid pursuits and goals in the modern world. Every single one of these items and scores of others that make up the operating assumptions of the age of growth are inimical to the principles of ecology, a low-entropy economic framework, and, most importantly, the newly defined stewardship doctrine.

Stewardship requires that humankind respect and conserve the natural workings of God's order. The natural order works on the principles of diversity, interdependence, and decentralization. Maintenance replaces the notion of progress, stewardship replaces ownership, and nurturing replaces

256

engineering. Biological limits to both production and consumption are acknowledged, the principle of balanced distribution is accepted and the concept of wholeness becomes the essential guideline for measuring all relationships and phenomena. In reality, the new stewardship doctrine represents a fundamental shift in humanity's frame of reference. It establishes a new set of governing principles for how human beings should behave and act in the world.

If the Christian community fails to embrace the concept of a New Covenant vision of stewardship, it is possible that the emerging religious fervour could be taken over and ruthlessly exploited by right-wing and corporate interests. The evangelical awakening could end up providing the essential cultural backdrop that a fascist movement in the United States would require to maintain control over the country during a period of long-range economic decline.

Even a thoughtful and respected evangelical theologian of the stature of Francis Schaeffer believes that fascism is a very real possibility for the United States in the troubled economic years that lie ahead. In reflecting on America's inability to find a solution to the problem of worsening inflation and recession cycles, Schaeffer concludes: 'I cannot get out of my mind the uncomfortable parallel to the Germans' loss of confidence in the Weimar Republic just before Hitler, which was caused by unacceptable inflation. History indicates that at a certain point of economic breakdown people cease being concerned with individual liberties and are ready to accept regimentation.'[57]

Schaeffer is pessimistic about the prospects for the United States. He believes that the overriding value Americans place on their own 'personal peace and affluence' will likely lead to a fascist-type order as the economy continues to contract: 'I believe the majority ... will sustain the loss of liberties without raising their voices as long as their own lifestyles are not threatened.'[58]

What Schaeffer fails to say is that there are already many disturbing signs within the evangelical movement pointing to just such a possibility. For example, many middle-class Christians are falling back more and more on the old notion

257

of the 'gospel of wealth,' equating Biblical doctrine with rugged individualism, free enterprise, and unlimited material accumulation. This kind of expansionist theology is still very much a dominant motif in American Christianity. The 'gospel of wealth' theme will likely continue to be used by individual Christians to justify a lack of concern or involvement with the pressing economic needs ahead, needs which require a communal and not merely an individual or free-enterprise response. For these Christians, the evangelical movement will serve as a sanctuary for withdrawal from the turmoil around them. If economic conditions become so bad that they begin to threaten even this last refuge of the middle class, chances are good that withdrawal will quickly translate into active support of right-wing and capitalist interests even to the point of accepting whatever authoritarian measures are deemed necessary by the state to maintain social order.

By radically redefining humanity's relationship to the rest of God's creation, contemporary Christian scholars are thrusting a theological dagger directly into the heart of the expansionist epoch. The new concept of dominion as stewardship and conservation rather than ownership and exploitation is at loggerheads with both traditional Christian theology and the mechanical world view of the past several hundred years. By refocusing the story of Creation and humanity's purpose in the world, Christian theologians have committed an act of open rebellion against their own doctrinal past. The Christian individual who for hundreds of years sought salvation through productivity and subduing of nature is now being challenged by a new Christian person who seeks salvation by conserving and protecting God's creation. *The Christian work ethic is being replaced by the Christian conservation ethic.* This new emphasis on stewardship is providing the foundation for the emergence of a new Christian Reformation and a New Covenant vision for society.

Facing the Entropy Crisis

There is no way to escape the Entropy Law. This supreme physical rule of the universe pervades every facet of our existence. Because everything is energy, and because energy is irrevocably moving along a one-way path from usable to nonusable forms, the Entropy Law provides the framework for all human activity. As we have seen, the entropy world view challenges our most treasured, and commonplace, assumptions about our environment, our culture, our very biological being. The trappings of modern culture – our great urban areas, our mechanized agriculture, our massive production and consumption, our weapons, our education, and our medical technologies – are all revealed in a radical new light. The Entropy Law shatters our view of material progress. It reorients the very foundation of economics. It transforms the notion of time and culture, and strips technology of its mystique.

Once we begin to understand the vast social and economic implications of the second law of thermodynamics, we come to understand that our existing world view bears absolutely no relationship to the way the world actually works. Our daily lives – our work, our play, our consumption, our very thoughts – lose their certainty, their grounding. We become strangers in a strange land. All of a sudden what used to be a clear and solid reality becomes fantasy, no better contrived than the Wonderland of the looking-glass world visited by Alice.

And yet, we resist the new orientation placed upon the world and our lives. Even as we are lured by the wisdom that emanates from an entropic world view, we struggle to keep our minds from being subverted by a vision whose import we can scarcely fathom. This is only natural, for we are being challenged to discard the safe and familiar myths that govern our existence. For many, of course, the prevailing myths have already lost their allure. Millions of Americans, some out of choice, others of necessity, are already adopting bits and pieces of the low-entropy philosophy and life-style. Increasingly, high-entropy concepts such as 'material progress at any cost' and 'bigger is better' no longer command the allegiance of as many inhabitants of the modern technological state as they once did. Some of these alienated heirs of the Newtonian world view will thus naturally welcome the liberation that comes from a shift in reality toward the entropic world view.

At the same time, it is also true that many others will struggle to deny the coming of the new age, preferring decaying familiarity to uncharted opportunities. Trapped within the framework of a philosophy they barely comprehend, these people will turn their attention to finding some mechanism that will provide them with an out. This, too, is only natural. We have been trained to think that there is always an out, that no force is beyond the human's ability to manipulate. We have been taught that there are no limits, that only narrow minds that have lost their nerve will give in to limits. But twist and turn as we might, there is no out.

In some ways we are like the man who refuses to believe in gravity. To prove its nonexistence – or at least his ability to overcome it – he climbs to the top of a large skyscraper and jumps. Gravity, of course, couldn't care less whether the man believes in it or not, and so it proceeds to administer a lesson to the sceptic by pulling him inexorably toward the ground. But the man, grasping at any straw to preserve his intellectual and physical survival, hurtles past the fortieth floor, proclaiming, 'So far, so good.'

If we, like the man who denies gravity, choose to deny the consequences of the Entropy Law, then we too will be taught an ultimate and shattering lesson. And, no doubt, like that man, we will continue to say, 'So far, so good,' even as the world around us disintegrates into chaos as a result of our high-entropy culture. Already we can anticipate at least three generalized responses from those who cannot bring themselves to cast off the prevailing world view.

First, there will be the *optimists*. They will pin their hopes on the assumption that somewhere, just over the next hill, or in the next laboratory, a technological solution will come along that will allow us to continue in our ways. Their faith soundly rooted in the values of modern society and the benefits of progress, they will unite under the well-worn banner of 'there's always a way.' They will tell us that 'you can't stop progress,' and that 'the American standard of living is the envy of the world.' Tying these hackneyed phrases to the assumption that the more material wealth a society has, the better off it is, these people will seek any possible measure to overcome our planetary limits.

The optimists are likely to concentrate their efforts on finding new ways to exploit renewable energy sources. While there is no question that we are about to shift from an energy base of nonrenewables back to renewables, what still remains in doubt is the type of energy transformers and flow line that will be set up. The technological optimists reject the notion of returning to a low-entropy flow and a greater accord with the natural rhythms and processes of the earth's ecosystems. Instead, they are placing their hopes on new genetic engineering technologies which they say will enable us to speed up the process of biological evolution and provide us with an increasing flow-through of matter-energy. If we are running out of petrochemical-based fertilizers necessary for mechanized agriculture, then we will develop a genetic engineering technique to construct plants that can fix their own nitrogen directly from the air. If oil is running out, we will genetically engineer and then mass-produce microorganisms that can substitute for depleted nonrenewable stocks.

It is even likely that the optimist will advocate the 'ordering' of the individual's biology. It is no coincidence that genetic engineering is moving out of the laboratories and into the region of applied science at this moment in our history. As entropy builds, our bodies internalize the disorders in the form of cancer, birth defects, diminished IQ in infants, and so on. The technological optimist, realizing that such disorders can seriously affect the nation's desire and ability to continue ever-greater economic growth, thus seeks solutions in bioengineering. If radiation and synthetic organic chemicals cause cancer and birth defects, we are already being told, then modern technology will cure us by rearranging our genes.

261

With genetic engineering the concept of high-energy, high-performance industrial production will be integrated directly into our bodies as people are artificially produced to standardized, technological specifications. In his never-ending quest for efficiency in all things, the optimist will likely seek to make life itself more 'biologically efficient.'

The optimists argue that we are not only moving from the age of nonrenewables to the age of renewables, but equally from the age of physics to the age of molecular biology. They point to the incredible scientific breakthroughs in genetic engineering in recent years and claim that within the next two decades our existing industrial technostructure will begin to give way to an entirely new set of technological transformers derived from bioengineering. Just as applied physics was used to transform a nonrenewable energy base into the accoutrements of the Industrial Age, applied genetic engineering will now transform a renewable energy base into an entirely new way of life, the biotechnical age.

It is interesting to observe that as the system attempts to shift from a nonrenewable energy base to a renewable one, and from applied physics as the transforming process to applied molecular biology, a new scientific paradigm is likewise emerging, one that the optimists hope will provide the basis for a new world view for the genetic age they are preparing for. The paradigm is called the *theory of dissipative structures* and its chief architect is Ilya Prigogine, the Belgian physical chemist whose work in nonequilibrium thermodynamics won him the Nobel Prize in chemistry in 1977. *Dissipative structures* refers to open systems that exchange energy with their environment. All living things, and some nonliving systems, are dissipative structures. They maintain their structure by the continual flow of available energy through their system. Prigogine points out that the more complex the dissipative structure, the more integrated and connected it is and thus the more energy flow-through it requires to maintain itself. Noting that the flow of energy through a dissipative structure causes fluctuation, Prigogine rightfully concludes that if the fluctuations become too great for the system to absorb, it will be forced to reorganize. Prigogine then asserts

262

that the reorganization always tends toward a higher order of complexity, integration, and connectedness and greater energy flow-through. Each successive reordering, because it is more complex than the one preceding it, is even more vulnerable to fluctuations and reordering. Thus, complexity creates the condition for greater reordering and a speedup of evolutionary development and energy flow-through. Prigogine, then, equates instability with flexibility. Using complicated mathematical formulas, he attempts to show that the more complex and energy-consuming the system is, the more flexible it is and the better able it is to change and readapt to new circumstances.

No matter that this theory flies in the face of our everyday common sense. We experience a world where increased complexity is narrowing our options, creating greater inflexibility, and increasing the likelihood of collapse and fragmentation. The theory of dissipative structures is an attempt to provide a growth paradigm for an energy environment based on renewables, just as Newtonian physics provided a growth paradigm for a nonrenewable energy environment.

It should be remembered that Newtonian physics was tailormade for 'nonliving' energy resources. It deals with dead matter in motion, with pure quantity. Therefore, it is an entirely inappropriate paradigm for an energy environment that is alive, renewable, and flowing. In contrast, the theory of dissipative structures provides a convenient scientific basis for the manipulation of 'living' energy sources, and it is for that reason that it is being heralded as a revolutionary breakthrough commensurate in scale with Newton's laws. As an overarching paradigm, the theory of dissipative structures provides a perfect rationalization for the age of bioengineering. It places a positive value on increased biological complexity and the continued reordering of living matter into new structures, which is what genetic engineering is really all about. With dissipative structures we move from viewing the world as an industrial machine to viewing it as an engineered organism.

In the next few years there will be a mad scramble to embrace renewable resources as the new energy base, genetic engineering as a new technological transformer, and the theory of dissipative structures as the new scientific paradigm. Greater energy flow-through, unlimited growth, and material progress without end will continue to dominate the thinking of those in power.

In an effort to ignore the Entropy Law, the experts will attempt to convince the rest of us that with a renewable energy base we will never run out of resources, and that growth will go on forever. In the short run, new genetic technologies, like recombinant DNA, might greatly increase the matter-energy flowing through the system, just as the first industrial transformers did with nonrenewables. For a time at least, it may well appear that we have overcome the fixed limits of the earth's ecosystem. That time span will be short-lived. In terms of its intimate effect on our day-to-day lives, the age of physics lasted less than a hundred years. If we proceed into the age of molecular biology, we can expect the span from beginning to end to be greatly reduced: the entire age may run its course in less than half a century. That's because the increased flow of matter-energy through the system will create disorders of an even greater magnitude than those produced by the massive flow of nonrenewable energy through the system.

First, by 'hot-wiring' the flow of living matter through society's energy flow line, we deplete in absolute terms the available stock of living matter. In a literal sense, renewable resources are really nonrenewable. That is, while they continue to reproduce, each blade of grass or microorganism produced today means one less in the future. In the words of Georgescu-Roegen, 'Matter, matters.' While the solar flow is virtually unlimited, the matter-energy that makes up the earth's crust is not. The earth's matter is continually degrading and dissipating. Natural recycling only reclaims for future use a part of whatever matter-energy is used up. The rest is irretrievably lost. Thus, the faster we speed up the flow of matter-energy through the system, the faster we will run out of renewable resources, regardless of how long the sun shines.

At the same time, the effect that the escalating entropy is

likely to have on the gene pool and the earth's fragile ecosystem could well be cataclysmic, doing much greater damage to the planet than was wrought during the entire age of nonrenewable energy flow.

The theory of dissipative structures, like the earlier Newtonian paradigm, completely ignores the Entropy Law, concentrating only on that part of the unfolding process that creates increasing order. By refusing to recognize that increased ordering and energy flow-through always creates ever greater disorder in the surrounding environment, those who advocate bioengineering technology as the transforming apparatus for a renewable energy environment are doomed to repeat the same folly that has led us to the final collapse of our nonrenewable energy environment and the age of physics that was built upon it.

As entropy, in its various forms, continues its dramatic rise within the high-energy culture so gloried in by the optimist, it will become necessary to attempt to maintain rigid order amidst the developing chaos. A true believer in the existing world order, the optimist will more and more condone practices and techniques that will become increasingly repressive and dehumanizing. For example, unable to admit that the high-entropy, megamillion city is simply not a viable living pattern, he will likely support the imposition of whatever police-state techniques are deemed necessary to maintain social order. Already, television cameras are appearing on street corners, satellites spy on us from outer space, and 'criminal' minds are electronically shocked into passiveness. Similarly, to maintain our nation's role as the planet's chief energy user, the optimist will encourage higher defence budgets and further weapons development in an attempt to protect a dwindling empire.

Of course, all this activity is doomed to failure. Each attempt at forcing order with new high-energy technologies will only speed up the chaos. Genes will be manipulated to create new forms of renewable energy or to cure disease or to raise IQ, but in the process, the evolutionary wisdom of billions of years will be irreversibly destroyed. Attempts to stem rising social disorders such as crime with new high-energy surveillance and weapons technologies will drain off precious energy from the rest of society and only create new forms of repression and antisocial behaviour. The optimist will not be able to win in his megalomaniacal quest for order, but he may succeed in taking all of humanity down with him.

The second general response to the Entropy Law can be called the *pragmatic*. Less of a true believer than the optimist, and far less grandiose in his schemes, the pragmatist will attempt to tinker with the existing structure in an effort to make it reflect at least some of the implications of the entropic world view. By nature, the pragmatist has a limited view of the world. He will comprehend part of the entropic paradigm but will miss its overall import. He will be willing to admit some of the shortcomings of the present system, but, after all, he will continue, that's the way the world is. New York City won't just go away; we can't sustain urban life unless we have mechanized agriculture and food processing; Americans will never give up their love affair with the automobile. Let's be realistic, he will say.

Of course, being a pragmatist, he will not deny that there is considerable room for improvement. 'Let's get more out of less' becomes his motto. Fine-tuning the existing high-energy structure will become his lifelong occupation. City planners will busy themselves developing thermodynamically sound transportation systems, well-insulated buildings, and neighbourhood advisory councils to encourage greater conservation. Auto makers will give us more miles to the gallon and cars that run on gasohol or electricity. Politicians will herald the virtues of 'lowered expectations' and 'planetary realism.' They will be careful, of course, to leave the overall techno-structure in place.

Even the most established powers will become engrossed with fitting their institutional imperatives into some kind of loosely defined entropy framework. In August of 1979, for instance, the Department of Energy sponsored a three-day conference on the second law of thermodynamics. Papers at the conference carried such weighty titles as 'Thermo-dynamic Analysis of Energy Efficiency in Catalytic Reforming' and 'The Reduction of Product Yield in Chemical Process by Second Law Penalties' and 'A Second-Law Taxonomy of Combustion Processes.'

In the future, we can no doubt expect heated technical debate from the pragmatists over the question, 'Given that entropy is always rising, what rate of increase is acceptable?' Theirs will be an effort at quantification, an attempt to take existing systems and make them most 'efficient' in the narrowest of senses. Thus, a world view will be converted in their minds into another cost-benefit tool. Unable to grasp the supreme reality of the Entropy Law, the pragmatist will miss the point entirely. Rather like a Christian asking, 'How much sin can I get away with and still make it to Heaven?' the pragmatist will be very good at adopting bits and pieces of the entropy vocabulary, all the time missing the essential message of the Entropy Law.

This is not to dismiss entirely the value of thermodynamic systems analysis. But before we can arrive at a point where such analysis has real meaning, we must first recognize that the Entropy Law tells us that a society's energy flow must be reduced to as low a point as possible in order to sustain the unfolding of all of life as far into the future as possible. The entropy economy is one of necessities, not luxuries. Once this is understood, then a basis is established that allows us to selectively use thermodynamic concepts as a tool to help organize the low-entropy society.

Compare, for example, the way a thermodynamic pragmatist (of the fine-tuning, get-more-from-less school) would tackle a set of problems, as opposed to the way these problems would be viewed by someone who has fully internalized the overarching importance of the Entropy Law. A pragmatist would look at an automobile and ask questions such as, 'How do we use the second law to redesign the motor so that we can get more work out of it?' and, 'What is the most thermo-dynamically sound design for a car's body?' A person who has fully comprehended the entropic world view would ask a far different set of questions. These might include: Are auto-mobiles really necessary for sustaining human life? Does the automobile enhance our well-being, our health, our culture?

Do today's automobiles rob succeeding generations of their own ability to sustain life?

The person who has internalized the low-entropy world view will always ask these comprehensive questions before getting down to specifics. He will understand that if something is not worth doing in the first place, then it really doesn't matter whether or not it is done well. If automobiles aren't worth having, then it makes very little difference whether we have cars that get twenty or fifty miles to the gallon.

What the pragmatist cannot comprehend is that the Entropy Law is the ultimate scientific law governing the physical world, not a tool that can be used to patch up the old system. The pragmatist does not understand that the Entropy Law completely redefines even the most familiar concepts of our everyday lives. The second law of thermodynamics shows us, for instance, that time is a function of entropy. When the world reaches a maximum state of entropy, and no more energy is available to perform work, time will cease, for nothing will be taking place. In an entropic sense, the only way to 'save' time is to keep a society's energy flow as close as possible to that which naturally takes place in our environment. In this way, the end of time and life will approach as slowly as possible. But the pragmatist will try to 'save' time by attempting to streamline the existing energy flow. This will only escalate the entropy process and, along with it, decrease the amount of time available to sustain life for generations yet to come. Similarly, the pragmatist will fail to comprehend the phenomenon of economic 'growth.' The pragmatist will turn his attention to defining the 'right' kind of growth, not realizing that the Entropy Law reveals to us that 'growth' is really a decrease in the world's wealth, nothing more than a process to take usable energy and transform it into an unusable state. Entropy shows us that the more an economy grows, the more it digs itself into a hole.

The third type of response to the Entropy Law might be called the *hedonistic*. Their motto is, 'Let's go out with a bang!' with the subtheme, 'What has posterity ever done for me?' These individuals are likely to agree that, in an overall sense, things are indeed getting worse. They will complain about air pollution, poisons in their food, the destruction of open spaces. But in a kind of last-days-of-Rome syndrome, they will argue that there's really nothing that can be done. Human nature is simply greedy and destructive, they will claim. Every time someone tries to change the system, they will point out, things just stay the same, unless they get even worse. What can the little person do except look out for number one; eat, drink, and be merry; and wait for the approaching doom?

The optimist, the pragmatist, and the hedonist all share one thing in common – a belief that as bad as things might be today, our generation still has a higher understanding of and control over reality than any previous generation. They each regard humanity before the modern age as being little more enlightened than beasts of burden. Because they did not know about subatomic particles, computers, and stereos, our ancestors must have been less human than we are now. These stalwart supporters of the Newtonian paradigm don't understand that we simply possess a different *kind* of knowledge than that which people held 500 or 5,000 years ago. From our reductionist viewpoint, we seem to know more and more. At the same time, we seem to comprehend less and less of what is happening to us. Completely divorced from nature, our urbanized intellects really have no insight into our true relationship to our environment. Our high-energy culture has, in fact, so fragmented our minds that we are no longer in harmony with the source of life. Divorced as we are from nature, we have no real chance to become enlightened, as that word has been understood by peoples throughout history. True, our ancestors had no scientific understanding of and explanation for the phenomena around them, but perhaps they had a better intuitive grasp of what was really important in life.

Our ancestors, at least, were self-sufficient. They knew how to provide for their needs. We, on the other hand, are the complete captives of our high-energy environment. We cannot grow our own food, provide our own entertainment, clothe ourselves with our own hands. We are like helpless infants whose every need must be serviced. In a marvellous passage, Wendell Berry, farmer and author, portrays our modern dilemma:

269

[An American] is probably the most unhappy citizen in the history of the world. He has not the power to provide himself with anything but money, and his money is inflating like a balloon and drifting away, subject to historical circumstances and the power of other people. From morning to night, he does not touch anything that he has produced himself, in which he can take pride. For all his leisure and recreation, he feel bad, he looks bad, he is overweight, his health is poor. His air, water, and food are all known to contain poisons. There is a fair chance that he will die of suffocation. He suspects that his love life is not as fulfilling as other people's. He wishes that he had been born sooner, or later. He does not know why his children are the way they are. He does not understand what they say. He does not care much and does not know why he does not care. He does not know what his wife wants or what he wants. Certain advertisements and pictures in magazines make him suspect that he is basically unattractive. He feels that all his possessions are under the threat of pillage. He does not know what he would do if he lost his job, if the economy failed, if the utility companies failed, if the police went on strike, if the truckers went on strike, if his wife left him, if his children ran away, if he should be found to be incurably ill. And for these anxieties, of course, he consults certified experts, who in turn consult certified experts about *their* anxieties.

Berry concludes, 'In living in the world by his own will and skill, the stupidest peasant or tribesman is more competent than the most intelligent workers or technicians or intellectuals in a society of specialists.'[59]

From Despair to Hope

Our generation faces a rare moment in human history. As this book has argued time and again, the energy environment influences the culture, values, politics, and economics that a society establishes. Now that we are witnessing the transition from an energy environment built upon nonrenewable resources to one built on solar flow and renewable energy sources, great personal and institutional changes will sweep over our society. The questions that confront us are: How long will the transition take? How will it be accomplished? What will be our individual roles?

The question of timing is the most difficult to speculate upon. The energy crisis that began in the early 1970s, and the public's concern about the destruction of the environment, have already set the stage for the rudimentary emergence of the entropic paradigm. No doubt, in the decades to come, bits and pieces of the entropic society will continue to develop, even as many vestiges of the old order linger on. Much the same process unfolded in Europe during the transition period between the medieval and the modern eras. Even today, visitors to European countries can observe remnants of feudal culture perpetuating themselves centuries after the disappearance of the system. In this sense, the transition to the Solar Age will likewise be an evolutionary development that will gain added momentum with each new entropy crisis.

At the same time, we should not delude ourselves into believing that the process of changing energy environments will be so gradual that business-as-usual will continue with only minor disruptions. The transition period will not extend over hundreds of years, as happened during the shift of previous energy environments. Our high-energy social and economic system is so fragile, so absolutely dependent upon continued inputs of nonrenewable resources, that monumental collapse could come at any time. Certainly, we can anticipate that the next twenty to thirty years represent the key period in launching the shift in energy environments. For this reason, we must begin to prepare, now, to minimize the possible shock waves that will naturally occur during this period of the entropy watershed.

A previous section outlined some of the broad, long-term institutional changes that will accompany the shift in energy environments. Some might think such changes utopian (and impossible to achieve); others might well regard them as oppressive (and therefore undesirable). To both groups it can only be said that if the low-energy future that has been outlined is unachievable or undesirable, what is the alternative? The scarcity of nonrenewable resources makes it clear that we can no longer maintain our existing high-energy industrial infrastructure. As we move from a nonrenewable energy base to a renewable one, it is equally unrealistic to expect that we can support, for very long, a continuing high matter-energy flow-through with bioengineering technology.

Like it or not, we are irrevocably headed toward a low-energy society. It is up to us whether we get there because we want to, because we understand both the necessity for our own survival and the vast opportunities for a better existence, or whether we try desperately to hang onto our existing world view and, finally, painfully, are forced into the future.

For every day that we continue along the present high-energy course, we are adding to the entropy bill that must eventually be paid. The shift to a whole new economic infrastructure derived from genetic technology might postpone the day of reckoning, but then only for a brief moment of time. The longer we put off the necessary transition from a high- to a low-entropy society, the bigger the entropy bill becomes and the more difficult the turnaround becomes. If we wait too long, we will find that the price that must be paid is beyond the ability of the human race to absorb.

The alternative to this wholesale squandering of available energy is an internalization of the values and dictates of the entropic paradigm. Unless we, individually and in unity with others, discard our Newtonian world view, there is no hope that a movement will develop that can revolutionize our society. The first step in this historic process is to fully comprehend what it is, as people, we believe. We must voluntarily reformulate our lives so that they reflect the new paradigm. But that is not enough. We must also join together, in a popular, grassroots social force, to begin the dismantling of the existing high-energy infrastructure. At the same time, we must build our new society based on a new set of values which reflect our awareness of the entropy process.

Perhaps all of this defies imagination. The task seems so great, the possibility of success so small. Once having come in contact with the Entropy Law, many will feel there is no hope. At first, the new world vision may seem profoundly depressing, and they will be left with nothing but despair. Where is hope? How can anyone be hopeful for a better future when, no matter what we do, we will leave the world in a more degraded form than it was when we were born? Where is hope when it appears that almost everything humanity has done in the past several hundred years has had a result exactly opposite to what was intended?

If we continue to base our hope on maintaining the existing order, then, truly, we will have only despair for our companion, for there is no hope that the modern age as we know it can long continue. On the other hand, what is so desirable about even entertaining such a hope? Why should we hope for more complex technology and more wasteful economic growth, when it only serves to rob us of our future as a species? Continuing to have faith in our high-energy environment is not a hope but an illusion. We should not despair of relieving ourselves of this illusion. Rather, we should rejoice that our generation has the opportunity to begin a planetary transformation that will move our world from the brink of annihilation into a new order of the ages.

There is great beauty in the Entropy Law. It guides us through the cosmic theatre with a bittersweet authority, assured of the ultimate fate that lies ahead but leaving to us the decision of how to proceed.

Up to this point in history the human race has driven relentlessly forward, conquering everything in its path. Now that it has succeeded in capturing and exploiting virtually every major ecological niche on the planet, humanity finds itself at the crossroads of its own history. The colonizing mode is taking its toll. As humankind continues to try to maximize its energy flow-through, the world's total energy environment depletes faster and faster, and the dissipation and disorder mount to higher and higher levels. The only hope for the survival of the species is for the human race to abandon its aggression against the planet and seek to accommodate itself to the natural order.

Our transformation to a climactic mode, if it is to occur at all, must be the result of a conscious choice by the human race. The fact that we are now becoming aware of that choice means that we have the power to effect a decision. That awareness of choice comes from an understanding of the Entropy Law.

After a long, futile search to find out where we belong in the total scheme of things, the Entropy Law reveals to us a simple truth: that every single act that occurs in the world has been affected by everything that has come before it, just as it, in turn, will have an effect on everything that comes after. Thus, we are each a continuum, embodying in our presence everything that has preceded us, and representing in our own becoming all of the possibilities for everything that is to follow.

Because every event that ever was or will be is interconnected, we share an ultimate responsibility for the infinite past and future. What we do in this world reverberates into the remotest corner of the universe, affecting everything else that exists. How we choose to live our lives is not only our own individual concern. It is of concern to everything, because our actions touch everything.

The Entropy Law is pure art, and a concept to behold with wonder. Yet it strikes terror in most of us. We cannot accept the fact that our physical world will one day complete its journey and cease to exist, any more than we can accept the fact that our own individual sojourns on earth are of a fixed duration. The Entropy Law, however, tells us that every occurrence in the world is a unique experience; it is the uniqueness of every event that makes us aware of the respect we owe to everything that exists around us. The whole world is temporary. In its finiteness, we experience our own. In its vulnerability we experience our own. In its fragile nature we experience our own.

Yet we desperately search for immortality in this finite world while knowing there is none. There is a nihilism to our search. The finiteness of the world is a constant unpleasant reminder of our own. We tear into everything around us, devouring our fellow creatures and the earth's treasures even while telling ourselves that it is progress we are after. It is, in truth, our own immortality we seek. It's as if we were determined to destroy every last reminder of this finite world in the hope of ridding ourselves of the painful awareness of our own temporary nature. Our violent actions only bring us faster to our own demise and to the demise of the fixed endowment bequeathed to all future living things. Meanwhile, we remain unconcerned about the carnage and affliction because we believe that modern science and technology can develop a substitute for everything we use up in nature's storehouse.

Only when we learn to accept the finite nature of the world can we begin to appreciate how precious this gift called the earth really is. Only then will every occurrence take on a special meaning and will life itself be something worth cherishing and conserving. As Wilhelm Ostwald, the great philosopher and scientist, once remarked, 'The responsibility for every act has sense only if the act cannot be repeated, if what is done is done forever.'[60]

There are those among us who are willing to accept the finiteness of the physical world but who believe that the entropic flow is counterbalanced by an ever-expanding stream of psychic order. To these people, the becoming process of life is synonymous with the notion of an ever-growing consciousness. In the Newtonian scheme, human consciousness is perceived as moving on an uphill grade against the downward journey of the entropy flow. Eventually, it is believed, humanity's collective consciousness will expand to a point where it will escape the physical plane altogether, overcoming the Entropy Law in a kind of cosmic metamorphosis. Piercing through the physical veil of existence, the collective human consciousness will then begin a steady ascent into the ethereal world of spiritual enlightenment.

It is not hard to understand, then, why people also harbour the unstated belief that in a nongrowth or low-energy flow-through environment, consciousness will atrophy or be prevented from developing. The idea lingers on that consciousness must be constantly watered and cultivated by accelerated physical activity if it is to grow. Taken to its logical conclusion, this line of reasoning would suggest that greater energy flow-through and greater disorder and dissipation in the world create a more conducive environment for the nourishing of consciousness.

This just isn't so. Speeding up the physical flow doesn't insure greater spiritual development; quite the contrary. Transcendence comes out of quietude and the recognition of the beauty of 'being,' not out of discord and the travails of 'doing.' Hermann Hesse's Siddhartha had to sit down by the river and listen quietly to the flow in order to become one with it and to reach enlightenment. Human development up to now, however, has been bound up with resistance to the natural flow of things. The hallmark of the colonizing mode is the attempt to conquer and subdue. We continue to think of enlightenment as something to 'achieve' when it's really something to 'experience.' As long as we frantically struggle for enlightenment we will continue to resist the natural rhythm of the unfolding process and slide further away from the enlightenment we seek.

It should also be recognized that we often mistakenly associate new human ideas for organizing the physical world we live in with higher forms of consciousness. The two are not the same. In fact, social development and spiritual development have, for the most part, followed opposite trajectories throughout much of human history. They can only begin to converge once again when humanity surrenders its will to dominate and begins to adjust to a world not of our making but for which we were made.

We also make a mistake when we confuse the becoming process with progressing or evolving toward some future perfect state. We experience a rose becoming but don't perceive it as an imperfect predecessor of some more perfect flower that is likely to unfold sometime in the far-off future. Nor do we question the value of a particular rose existing. Its very existence is enough to justify it. The perfection of the rose is in its being. Why shouldn't it be the same with humanity? People have not changed in terms of physical and mental capacity for nearly two million years. Just as each rose is a rose and is therefore perfect in itself, that is, in its own subjective occurrence, so too with every human life.

It's ironic then, that we continue to hold on to the belief in the progressive unfolding of a collective human consciousness that will culminate in total enlightenment sometime in the far-off future, whereas in truth, the perfect state is ever present. Until we recognize that revelation and cosmic consciousness is available to everyone at all times, we will never accept full responsibility now for our every action and our relationship to the world around us. Instead, we will continue to rationalize our errors and omissions as being the result of our less than enlightened state in the collective becoming process. In other words, because we are not yet totally conscious, therefore, we do not yet have to be totally responsible.

Once we fully accept the Entropy Law, however, we can never again hide from our total responsibility for everything that happens in the world we live in and affect. Total responsibility, in turn, is a precursor to the experiencing of total consciousness and spiritual enlightenment.

The Entropy Law answers the central question that every culture throughout history has had to grapple with: How should human beings behave in the world? While it has been generally agreed that people should act in a way that preserves and enhances life, there have been countless prescriptions for exactly how to go about achieving such ends. Finally, the Entropy Law provides an answer that is all-embracing. Preserving and enhancing life, in all its forms, requires available energy. The more energy available, the greater the prospects for extending the possibilities of life into the future. But the second law also tells us that the available store of energy in the world is continually being depleted by every occurrence. The more energy each of us uses up, the less is available for all life that comes after us. The ultimate moral imperative, then, is to waste as little energy as possible. By so doing, we are expressing our love of life and our loving commitment to the continued unfolding of all of life.

Therefore, when we speak of love in the universal sense, we are speaking of that deep spirit of oneness that acknowledges that we are each an inseparable part of the total flow that is the becoming process of life itself.

Love is not antientropic, as some would like to believe. If love were antientropic, it would be a force in opposition to becoming, for the entropic flow and becoming go hand in hand. Rather, love is an act of supreme commitment to the unfolding process. That is why the highest form of love is self-sacrifice – the willingness to go without, even to give one's own life, if necessary, to foster life itself.

Love is a gentle, subtle force that conveys a feeling of total awareness and integration with the universal rhythm that is the becoming process. By its expression, love acknowledges a master plan for the unfolding of the physical sojourn in the universe, even as it acknowledges the impossibility of ever fully understanding the mysteries that lie behind it. It is at once a statement of faith in the ultimate goodness of that cosmic process and an act of total, unconditional surrender to the natural rhythmic flow that carries all physical reality along its course.

Love, then, is a savouring experience. It attempts neither to speed up nor to arrest the becoming process because in its pure form it is simply the embodiment of that universal cosmic rhythm that is meant to be respected and adhered to.

In the end, our individual presence rests forever in the collective soul of the unfolding process itself. To conserve as best we can the fixed endowment that was left to us, and to respect as best we can the natural rhythm that governs the becoming process, is to express our ultimate love for all life that preceded us and all life that will follow. To be aware of this dual responsibility is the first step toward our transformation from a colonizing to a climactic mode. We are the stewards of the world.

278

Postscript

There are seven things which every reader should bear in mind after reading through this essay. First, the earth is virtually a closed system. In thermodynamics, consideration is given to three types of systems: isolated systems which exchange neither matter nor energy with the outside world; closed systems which exchange energy but not matter; and open systems which exchange both matter and energy with the external environment. The earth is virtually a closed system in relation to the solar system. It exchanges energy with the sun, but for all practical purposes, does not exchange matter with the rest of the solar system. With the exception of an occasional meteorite that falls to earth, the spraying of small amounts of cosmic dust, and an occasional satellite sent into space, no significant amount of matter enters or leaves the earth.

Second, in the short run and in isolated geographic pockets around the planet, entropy watersheds are experienced. That is, the particular matter-energy base that a society is using becomes depleted, as a result of natural forces at work or as a result of people consuming resources faster than nature can reproduce them. This forces a change to a new matter-energy base. This essay in no way suggests that the final heat death of the planet is imminent. It does suggest that our current matter-energy base of fossil fuels and a combination of specific metals is becoming depleted, requiring the shift to a new matter-energy field.

Third, every new matter-energy base becomes the context for the development of a new set of technologies to collect, exchange and discard that particular matter-energy environment. Along with the new modes of technology, come new institutions, values, and world views. While the matter-energy base sets the context, it does not rigidly determine the specific ordering process that a society chooses to utilize in transforming the environment into the economic utilities of life. The types of technologies, institutions, values and world views can vary considerably, but they must at least remain compatible with the matter-energy base they are processing.

Fourth, the world economy is in the early stages of an historic transition from an extractive energy base of fossil fuels and rare metals to a solar age with renewable resources as the primary energy source. Already two competing methodologies are developing, each with a very different approach to organizing biological resources in the coming solar age. The first method can be loosely defined under the rubric of appropriate technology. This approach puts a premium on compatibility with the speed of the production process in nature. The overriding principle is to balance our economic budget with nature. In other words, an effort is made not to consume faster than nature can produce. Emphasis is placed on de-centralized institutions, labour-intensive skills, greater diversity and regional self-sufficiency along with frugal and equitable use of nature's resources. An appropriate technology approach and infrastructure is already developing in a fragmented fashion in communities across the country.

At the same time, an entirely different approach to organizing renewable resources is emerging as we enter the Solar Age. It's called genetic engineering. Many people mistakenly believe that genetic engineering is a technology. In a more profound sense, it is really a method for organizing a renewable resource base. Corporations are pouring billions of dollars into developing genetic engineering because they are beginning to recognize that the historic transition from fossil fuels to solar and renewable energy is upon us. Believing that the appropriate technology approach to organizing renewable resources is too slow and inefficient to maintain existing 'growth' patterns, it is argued that engineering the biology of the planet is essential in order to speed up the conversion of living matter beyond nature's own tempo, providing an ever expanding growth curve as we enter the Solar Age.

In the next two decades critical decisions will be made as to which of these two very different methods of organizing renewable resources will eventually dominate.

An understanding of the Entropy Law and the laws of thermodynamics is essential if we are to avoid the folly of a genetically engineered solar age. For a more detailed analysis of the grave danger of a genetically engineered solar age, the reader should refer to our book *Who Should Play God?* It examines the ecological, economic, political and moral issues raised by genetic engineering and the artificial creation of life.

Fifth, in a very long run, when the sun ultimately dies out, the earth will become a cold, barren planet and eventually dust swirling through the cosmic theatre. In the past, scholars have equated Entropy with the final heat death of the solar system and then concluded that it is not of great concern to human life since that eventuality is so far off into the distant future. *In contrast, this essay focuses attention on entropy as a process rather than a final state.* It examines the great shifts in matter-energy environments here on earth and the relationship of human beings to the laws of thermodynamics and the entropic flow. Its goal is to provide a framework for analysis. The political, cultural and economic struggles that unfold as people and civilizations adjust to a radical shift in their resource environment is examined, but not explored in depth. Hopefully, this emerging conceptual framework based on the laws of thermodynamics will encourage others to look at the political, cultural and economic dimensions of change with a new perspective.

Sixth, there will be those who find the Entropy Law utterly depressing. This is indeed strange since it is merely a physical law. When Copernicus announced that the universe does not revolve around the earth, many people were similarly depressed, but humanity somehow managed to adjust to reality. Physical laws merely tell us the way the physical world operates. How we choose to relate to those laws determines our frame of mind. It is curious to hear people lament that if the physical world is truly finite and is moving toward death with each passing moment, of what good is it to even try. Why not just give up? Yet our own personal lives also obey the Entropy Law. We go from birth to death. Our physical sojourn is finite and try as we will, there is simply no way to overcome that reality. As we first come to recognize our own finite existence, we don't generally say to ourselves, if it's all down hill (from birth to death) why bother trying at all. Instead, for most of us, the profound recognition of our own mortality usually propels us, for at least brief moments, to consider using every experience in life judiciously and with respect and reverence, knowing there is no substitute, alternative or reversal for anything we do in our personal lives. Unfortunately these moments of profound recognition of our own personal mortality are usually few and far between, the rest of our time taken up with the mad scramble to overcome the Entropy Law. What is true for our personal sojourn is equally true for the rest of the physical sojourn around us. Just as it is often difficult to accept our own physical mortality and the irreversibility of our life's experiences, it is difficult to accept the irreversible and finite nature of the world around us.

The entropy process is neither optimistic nor pessimistic. It is just a description of how the physical world unfolds. How we choose to come to terms with that process philosophically, determines our outlook individually and as a society. 'Coming to terms' means understanding that entropy itself is neither good nor evil. It is true that entropy represents decay and disorder, but at the same time also represents the unfolding of life itself. Values come to play when we make decisions as to how to interact with the entropic flow.

Finally, like all scientific constructs, Entropy and the laws of thermodynamics are anthropocentric in nature. All scientific laws represent our need to use symbolic abstractions to try and understand, as best we can, the way the physical world operates. If one were to ask what are the things we are most sure of in the physical world around us, at the top of the list would have to be the notion of birth to death, hot to cold, concentrated to dispersed, available to unavailable, value to waste, order to disorder, beginning to end. These ideas about the way the physical world unfolds are the embodiment of the Entropy Law.

Knowledge of the Entropy Law can help us understand our relationship to the physical sojourn of which we are a small, but significant part. Like gravity, entropy is only a physical law, and this should be understood by those who would either deny its relevance altogether or who would turn it into an all embracing ideology. As an anthropocentric concept, entropy can help define the physical rules within which the game of life unfolds. How that game is played, however, is determined by the values and visions, whims and caprices, ideologies and 'isms' that emanate from the human mind as people interact with each other and their environment.

Afterword

The history of thermodynamics has been – and still is – unusually agitated mainly because of the unique nature of the Entropy Law. Although the basic facts with which thermodynamics is concerned have very likely been known to humankind since the dawn of civilization, they were incorporated into the edifice of science only a hundred years ago. Before that time, men of science paid no attention to one of the most elementary facts, namely, that heat always passes by itself from the hotter to the colder body, never by itself in reverse. Today, this truth constitutes the most transparent formulation of the second law of thermodynamics, alias the Entropy Law.

The first interesting thing about this law is that it was established long before the other, less intriguing, laws – the first and the zeroth. The first law states that energy can be neither created nor annihilated (which implies the theoretically sensitive point that work of any kind is a form of energy). The zeroth law, the last to be added as a necessary theoretical pillar of classical thermodynamics, simply says that if two bodies are each in thermal equilibrium with a third, the two will also be in thermal equilibrium when brought in contact with each other.

Curiously, it was the acceptance of the first law that met with great difficulties. One can speculate that the human mind was somewhat reluctant to give up the hope that one day we may construct an engine that can perform work without using energy, that is, a perpetual motion of the first kind. One need only look up the first issue of *Science* to see that as late as 1880 – hence, after thermodynamics had become a legitimate branch of natural science – the belief that electricity represents an endless free source of motor power was going strong.

As far as the march of science is concerned, it was the Entropy Law that struck the first blow to the mechanistic dogma which reigned supreme ever since the spectacular successes achieved in astronomy by Newtonian mechanics. According to that dogma, processes can proceed both forward and backward, and, as Laplace emphasized in his celebrated apotheosis of mechanics, all in nature consists of simple qualityless motion. To the crisis, purists replied – and a few still do – that thermodynamics is not a legitimate natural science because some of its concepts are anthropomorphic (as if any human concept can have any other root). The Entropy Law indeed implies a distinction rooted in the structure of humans. One of its formulations calls for distinguishing between two qualities of the quantitatively invariable energy. There is the available energy, the quality that we can use for our own purposes, and the unavailable energy that, in the words of Lord Kelvin, is 'irrevocably lost to man . . . although not annihilated.' The point is a consequence of a principle first brought to light by N. L. S. Carnot (in a famous memoir of 1824): 'For a heat engine operating in cycles to perform mechanical work, we must use two bodies of different temperatures.'

Just as a weight can produce mechanical work only if it can fall from a higher to a lower level, so thermal energy cannot drive an engine operating in cycles unless it can 'fall' to a lower level of temperature (a parallel that Carnot wrongly interpreted as an analytical identity). Just as a weight cannot supply any mechanical work once it reaches the lowest available level, so thermal energy is 'irrevocably lost to man' after it reaches the lowest available temperature.

What is denied here is the perpetual motion of the second kind: that is, an engine operating in cycles and using only the thermal energy of a single source. But we must not fail to note that this denial would not apply if we were creatures that would not necessarily be confined to a finite space, so that we could also use engines that do not operate in cycles; as well as creatures for which time did not matter, so that we could use engines that, moving with infinitesimal speed, encounter no friction. The only reason that sets perpetual motion of the second kind out of bounds for us is the finitude of the human condition. Only on this basis can one say that thermodynamics smacks of anthropomorphism.

Because Rudolf Clausius defined entropy as a relative index (relative to temperature) of the unavailable energy in an isolated system, we speak now of the irrevocable increase of entropy, thus leaving the impression that the increase comes mysteriously from nowhere. What is overlooked is that the increase corresponds to a decrease in the available energy. For a simple representation of this otherwise complex phenomenon I liken the isolated system to an hourglass that cannot be turned upside down yet marks the true passage of time.

Because of the mysterious way the Entropy Law is usually formulated and because the great physicist A. S. Eddington hailed it as the supreme law of nature, that law has had an unusually strong appeal. The concept of entropy has also been transplanted into virtually all other domains – communications, biology, economics, sociology, psychology, political science, and even art. The culprit for opening the door for this situation is Claude Shannon. On finding in his path-breaking contribution of 1948 that the average number of messages per signal in a vernacular code is given by the same algebraic formula as that proposed by Boltzmann for entropy, Shannon referred to that average as 'the entropy of information.' The term has stuck ever since. A muddled semantic metamorphosis has then led even to the identification of knowledge with low (negative) entropy. But Shannon, at least, showed his scholarly stature by denouncing in his 1956 article 'The Bandwagon' the absurdity of the trend that has 'ballooned [the entropy information] to an importance beyond its actual accomplishments.' Not surprisingly, however, the parade with the naked emperor still marches on.

In view of the irreversibility proclaimed by the Entropy Law, it was quite natural that this law should create great stirring around the eternal issue: What is life? The establishment of the Entropy Law threw that issue onto the horns of a dilemma. If the material universe is constantly subject to irrevocable degradation, how can life-bearing structures develop, survive, and even expand? This thought, no doubt, prompted some of the great pioneers in thermodynamics to have reservations on the universal validity of the Entropy Law.

Quite early, Hermann von Helmholtz questioned whether the reversal of unavailable into available energy 'is also impossible for the delicate structures of the organic living tissues.' Still more interestingly, Lord Kelvin's first formulation of that law ran as follows: 'It is impossible *by means of inanimate material agency*, to derive mechanical effect from any portion of matter by cooling it below the temperature of the coldest of the surrounding objects' (emphasis added).

The open conflict sprang from an epistemological clash. Long before the Entropy Law was established, Karl Ernst von Baer (1792–1876) refuted the prevailing dogma that eggs are miniatures of developed creatures by discovering the mammalian ovum. He was thus led to proclaim that the heterogeneous emerges from the homogeneous. Later, Herbert Spencer raised this idea to the level of the most important law of nature. Still later, some scholars (such as George Hirth and Felix Auerbach) and philosophers (such as Henri Bergson and Alfred North Whitehead) insisted on the unique property of life to run counter to the downgrade movement of inert matter. The accusations of mysticism cast at this philosophy crumble if one simply observes the following:

First, the Entropy Law applies only to completely isolated systems, whereas a living organism, being an open system, exchanges both matter and energy with its environment. There is thus no contradiction to the Entropy Law as long as the increase in the entropy of the environment more than compensates for the decrease in the entropy of the organism.

Second, the Entropy Law does not determine the speed of the degradation; this may be accelerated (as by all animals) or slowed down (as by green plants).

Third, the same law does not constrain the types of structures that may emerge from the entropic whirlpool. For a clarifying analogy: geometry constrains the size of the diagonals in a square, but it does not constrain the colour of the square. To be sure, how a square happens to be 'green,' for instance, is a different and almost impossible question.

Thermodynamics teaches what Boltzmann and, quite recently, Erwin Schrödinger said, namely, that any organism needs to continuously suck low entropy from the environment; otherwise, it would very quickly degrade entropically. But no loophole has yet been discovered in the Entropy Law to justify the impressive claim that the existence of life-bearing structures is a necessary conclusion of thermodynamic laws. The truth as we know it today is that life is not a conclusion of the whole body of chemicophysical laws and that it is the *behaviour* of any chemical compound or biological organism that must be listed among the properties of every element composing that compound or organism.

This brief story has recently been completed with a very important episode. Either because the Western intellect is dominated by a flow complex or because energy, being a homogeneous 'substance,' is analytically far more tractable than heterogeneous matter, thermodynamics (or any other theoretical branch of physics for that matter) paid no attention to what happens to the material scaffold of engines. Matter is mentioned in thermodynamics only in relation to the waste of available energy through the nonuseful work against friction. Yet the fact that available matter (to retain Lord Kelvin's expressive term) also becomes unavailable is just as elementary and transparent as the similar transformation to which energy is subjected. That is not all. Transparent and elementary also is the fact that, again because of the finitude of our existence, we cannot recycle the rubber molecules dissipated from automobile tyres, the copper molecules dissipated from coins, the phosphorous molecules dissipated from chemical fertilizers, and so on down the line. These are 'irrevocably lost to man.' We can recycle only matter that is still available but is no longer in a useful shape: broken glass, worn-out tools, etc. – in a word, 'garbojunk.'

A new law, the fourth law of thermodynamics (not a very fortunate terminological choice) must, therefore, complete our description of the entropic transformations. It may be stated in several equivalent formulations:

A. Unavailable matter cannot be recycled.
B. A closed system (i.e., a system that cannot exchange matter with the environment) cannot perform work indefinitely at a constant rate.

This law proclaims for matter what the Entropy Law proclaims for energy. One difference is that in an isolated system, instead of tending toward heat death (when all energy is unavailable), it tends toward chaos (when all matter-energy is unavailable). However, we must refrain from speaking of the entropy of matter as a measurable entity. There is a measurable entropy for energy because energy is homogeneous; matter in bulk is instead heterogeneous, as is plainly evidenced by the Mendeleev table. The factors that dissipate matter, therefore, vary greatly from one material substance to another; hence, we cannot (at this time) subsume all material dissipations into one general formula – which does not mean that such dissipations do not take place irrevocably or that we cannot speak of the general degradation of available matter-energy into the unavailable form.

In this new light, the entropic predicament of our species emerges as far more complicated than we all now think in reacting to the present energy crisis. First, the recipe of a steady state can no longer be considered an ecological salvation (which does not mean to repudiate also the ethical and social merits that Herman Daly invokes in its support). Second, a valid technology must be able to maintain its material scaffold as long as its specific fuel is available. Considering only the energy flows, as is now the general practice, may mislead. Notwithstanding continuous claims to the contrary, the direct use of solar energy does not yet constitute a viable technology. The presently known recipes are certainly feasible (as the recipe for putting a man on the Moon is), but (like this other recipe) they are parasites of the current technology. Not to recognize this fact overtly fosters false and, hence, dangerous hopes in the public mind.

A still more interesting turn in the history of thermodynamics is the heated controversy, often exaggerated beyond reason, that grew around irreversibility; that is, around the unique direction of time as the stream of human consciousness. It was the peculiar attachment of our minds to the mechanistic explanation that burst to the surface again. As we may recall, in his Baltimore Lectures (1894), Lord Kelvin acknowledged that only if he was able to represent a process

289

by a mechanical model was he able to understand it. Naturally, the Entropy Law, which expressly denied that thermal energy can move by itself in both directions between two bodies, was not accepted wholeheartedly by the scholarly world. So, apart from a few protesting voices, all physicists were delighted when Boltzmann set forth the idea that thermodynamic phenomena are the result of the movements of the molecules of a gas according to the deterministic laws of classical mechanics combined with a random factor.

The mechanistic dogma thus triumphed again. That it has ever since remained the fundamental password in thermodynamics should not surprise us. But the flaws caused by the probabilistic viewpoint in our *Weltbild* and, finally, in our *Weltanschauung* should not pass unnoticed.

As long as the logical foundation of thermodynamics rests also on probability (in whatever formulation), the spontaneous turning of unavailable into available energy is only a very improbable but not impossible event. We can therefore hope to think up a sharper's trick by which to cause this possible transformation to happen almost at will and thus increase our supply of available energy. It is not unreasonable to think that this sanguine hope is responsible for the rather frequent belief that the Entropy Law will be refuted someday.

Unfortunately, this belief bears on mankind's entropic predicament mentioned earlier. It stems from the fact that the human species transcended the slow and uncertain biological pathway to ecological progress when it began producing exosomatic (detachable) organs out of mineral resources. The root of economic scarcity lies not only in the finitude of these resources but also in the irrevocable entropic degradation. It is for this reason that, fifteen years ago, in my analysis of the entropic nature of the material side of the economic process, I felt it necessary to expose in detail the fallacy of the marriage between probability and the strict laws of mechanics. That my effort was worthwhile finds proof in the outburst with which Peter L. Auer, a professional physicist, once prefaced his claim that the Entropy Law does not prevent continuous economic growth. This is the same position as that of the economic profession as a whole: 'Come what may, we shall find a way.'

It is rather hard to explain how this view was reached. True, the founders of neoclassical economics were infatuated with the mechanistic dogma dominant at that time. Economists have taken notice of the first law. Alfred Marshall, for one, explicitly recognized that we can produce neither matter nor energy; we can produce only 'utilities.' Modern economists, however, have failed to take notice of the Entropy Law; so none has come to ask how we can produce utilities. Briefly, standard economics (the economics prevalent nowadays) has completely ignored the special role of depletable natural resources in mankind's mode of life, a role that transpires through the main actions of history, especially through the history of warfare. Not only are depletable natural resources completely absent from standard economic theory, but only economic growth is 'the grand objective' of the economic science, as Sir Roy Harrod proudly proclaimed. Indeed, some of the greatest palms have been won by economic models in which continuous exponential growth is taken for granted. Naturally, to claim that to bring about growth is the economists' expertise constitutes the surest way to induce a general admiration for the profession.

After the oil embargo of 1973–1974 a few economists stealthily shifted their old positions somewhat. Walter Heller – an exception – even admitted that economists 'have been caught with their parameters down.' (It would have been more correct to say 'up in the air' than 'down.') Yet most economists stubbornly cling to the position that the price mechanism can prevent any scarcity calamity. The simple fact, noted by William Miernyk, is that the price of oil at the time when one could get free drinking glasses with each petrol fill-up pointed the United States technology and economy in a direction we all now regret. Because of the general growth mania and the economists' unchallenged faith in the price mechanism, some people now crave such utterly absurd gadgetry as the golf cart, while numberless others go through a very short life of sufferings beyond imagination.

Although the haves care, albeit marginally, for their contemporary have-nots, mankind as a whole does next to nothing to protect future generations from possible catastrophes.

291

Programmes such as my bioeconomic one have been variously proposed. The only reason why they have remained virtually ignored is the difficulty of changing values concerning intranational and, especially, international relations. It is therefore imperative for us all to realize the dangers for our whole species created by a behaviour based on individual self-interest and on maximizing personal utility instead of minimizing future regrets.

The Entropy Law in its extensive form sets material limits to the specific mode of life of the human species, limits that tie together present and future generations in an adventure without parallel in our knowledge. Because the importance of these limitations has come into plain view only recently and because the entropic abundance of the last two hundred years or so is rapidly approaching its end, we must reassess and remodel our approach to economic, political, and social evolution. Jeremy Rifkin is especially qualified to set this entire problem in a convincing light, not cluttered by insubstantial technical details. This volume is written with the same masterly human penetration that has won great acclaim for his earlier essays. In writing about the Entropy Law, one runs the risk of falling victim to the fashion of seeking to impress by complex but empty exercises. Jeremy Rifkin steered clear of the frequent exercises about the formal parallelism imagined to exist between entropic transformations and social phenomena: thermodynamics sets a limit to these phenomena but does not govern them. For its timely educative value and for its pronouncedly human underpinning this volume should have an honoured place on any individual or public bookshelf, to spread the commandment suggested by the present turning point in mankind's life on this planet: 'Love thy species as thyself!'

Nicholas Georgescu-Roegen
Vanderbilt University
February 1980

Notes

PART ONE
World Views

1. J. B. Bury, *The Idea of Progress: An Inquiry into Its Origin and Growth* (New York: Macmillan, 1932), pp. 11–12.
2. Pardon E. Tillinghast, *Approaches to History* (Englewood Cliffs, N. J.: Prentice-Hall, 1963), p. 9.
3. Ibid., p. 11.
4. John Herman Randall, *The Making of the Modern Mind* (Cambridge, Mass.: Houghton Mifflin, 1940), p. 34.
5. Dante Germino, *Modern Western Political Thought: Machiavelli to Marx* (Chicago: Rand McNally, 1972), p. 166.
6. Francis Bacon, *Novum Organum*, Book 1, Aphorism 2.
7. Ibid., Aphorism 71.
8. Randall, op. cit., p. 223.
9. Ibid., p. 224.
10. Theodore Roszak, *Where the Wasteland Ends* (Garden City, NY: Doubleday/Anchor Books, 1973), pp. 144–45.
11. Randall, op. cit., p. 224.
12. Jean Houston, 'Prometheus Rebound: An Inquiry into Technological Growth and Psychological Change,' in *Alternatives to Growth I*, Dennis Meadows, ed. (Cambridge, Mass.: Ballinger, 1977), p. 274.
13. Randall, op. cit., p. 241.
14. Ibid., pp. 241–42.
15. Ibid., p. 259.
16. Ibid.
17. Alfred North Whitehead, *Science and the Modern World* (New York: New American Library, 1925), p. 50.
18. Leo Strauss, *Natural Rights and History* (Chicago: University of Chicago Press, 1953), p. 258.

19. John Locke, 'Second Treatise,' in John Locke, *Two Treatises of Government*, ed. Peter Laslett (Cambridge University Press, 1967), p. 315.
20. Ibid.
21. Ibid.
22. Ibid., p. 312.
23. Ibid.
24. Ibid.
25. Adam Smith, *An Inquiry into the Nature and Causes of the Wealth of Nations*, ed. Edwin Cannon (London: Methuen, 1961), 1:475.

PART TWO
The Entropy Law

1. Theodore Roszak, *Where the Wasteland Ends* (Garden City, NY: Doubleday/Anchor Books, 1973), p. 139.
2. Isaac Asimov, 'In the Game of Energy and Thermodynamics You Can't Even Break Even,' *Smithsonian*, August 1970, p. 9.
3. Ibid., p. 6.
4. Herman Daly, *Steady-State Economics* (San Francisco: Freeman, 1977), pp. 21–22.
5. Nicholas Georgescu-Roegen, 'The Steady State and Ecological Salvation,' *Bio Science* (April 1977): 268.
6. Ibid.
7. Bertrand Russell, *The Scientific Outlook* (New York: Norton, 1962), p. 114.
8. Stanley Angrist and Loren Hepler, 'Demons, Poetry and Life: A Thermodynamic View,' *Texas Quarterly* 10 (September 1967): 27–28.
9. Ibid.
10. Ibid., p. 29.
11. Russell, op cit., pp. 90–91.
12. Nicholas Georgescu-Roegen, *The Entropy Law and the Economic Process* (Cambridge, Mass.: Harvard University Press, 1971), pp. 6–7.
13. Angrist and Hepler, op. cit., p. 30.

14. G. Tyler Miller, Jr., *Energetics, Kinetics and Life* (Belmont, Calif.: Wadsworth, 1971), p. 46.

15. Philip P. Weiner, ed., *Dictionary of the History of Ideas* (New York: Scribner's, 1973), 2:113.

16. P. A. Y. Gunter, ed., *Bergson and the Evolution of Physics* (Knoxville, Tenn.: University of Tennessee Press, 1977), p. 63.

17. William Thompson (Lord Kelvin) in *Proceedings of the Royal Society of Edinburgh* 8:325–31 (1874). Cited in *American Scientist*, October 1949, p. 559.

18. Harold F. Blum, *Time's Arrow and Evolution* (Princeton, N.J.: Princeton University Press, 1968), p. 94.

19. Erwin Schrodinger, *What is Life?* (New York: Macmillan, 1947), pp. 72, 75.

20. Leslie A. White, 'Tools, Techniques and Energy,' in *Cultural and Social Anthropology*, ed. D. Hammond (New York: Macmillan, 1964), p. 28.

21. Miller, op. cit., p. 291.

22. Ibid.

23. White, op. cit., p. 28.

24. Alfred J. Lotka, 'Contribution to the Energetics of Evolution,' *Proceedings of the National Academy of Science* (1922): 8:149. Also see, Lotka, 'The Law of Evolution as a Maximum Principle,' *Human Biology* (September 1945): 186.

PART THREE
Entropy: A New Historical Frame

1. Harry Rothman, *Murderous Providence: A Study of Pollution in Industrial Societies* (New York: Bobbs-Merrill, 1972), p. 34.

2. Lynn White, Jr., 'Technology in the Middle Ages,' in *Technology in Western Civilization*, ed. Melvin Kranzberg and Carroll W. Purrell, Jr. (New York: Oxford University Press, 1967), p. 72.

3. William McNeill, *Plagues and People* (New York: Doubleday/Anchor Books, 1976), p. 147.

4. Lewis Mumford, *Technics and Civilization* (New York: Harcourt, Brace, 1934), pp. 119–20.

5. Ibid., p. 120.

6. Eugene Ayres, 'The Age of Fossil Fuels,' in *Man's Role in Changing the Face of the Earth*, ed. William L. Thomas, Jr. (Chicago: University of Chicago Press, 1956), p. 368.

7. Edmund Howes, ed., *Stow's Annals* (London, 1631), quoted in W. H. G. Armytage, *A Social History of Engineering* (London: 1961).

8. Richard Wilkinson, *Poverty and Progress* (New York: Praeger, 1973), pp. 90, 102.

9. Friedrich Engels, *The Condition of the Working Class in England* (Oxford University Press, 1958), pp. 78–79.

10. Jacques Ellul, *The Technological Society* (New York: Random House/Vintage Books, 1964), p. 105.

11. Ibid., p. 116.

12. Eugene S. Schwartz, *Overskill: The Decline of Technology in Modern Civilization* (New York: Quadrangle, 1971), p. 72.

13. 'Innovation: Has America Lost Its Edge?' *Newsweek,* June 4, 1979, pp. 58–59.

14. *Environmental Quality*, Ninth Annual Report of the Council on Environmental Quality (Washington, DC: US Government Printing Office, 1978), p. 437.

15. 'Will Industry Come to Terms with Mother Earth?' *Industry Week,* February 5, 1979, p. 70.

16. Information available from the National Petroleum Institute.

17. 'Innovation,' ibid., p. 59.

PART FOUR
Nonrenewable Energy and the Approaching Entropy Watershed

1. Fred C. Allvine and Fred A. Tarpley, Jr., 'The New State of the Economy: The Challenging Prospect,' in *US Economic Growth from 1976 to 1986: Prospects, Problems and Patterns,* Studies for the Joint Economic Committee of the US Congress (Washington, DC: US Government Printing Office, 1976), p. 58.

2. Wilson Clark, *Energy for Survival* (Garden City, NY: Doubleday/Anchor Books, 1975), p. 70.

3. Resources for the Future, *Annual Report*, Washington, DC: 1972, p. 12.

4. Luman H. Long, ed., *The 1972 World Almanac* (New York: Newspaper Enterprise Association, 1971), p. 456.

5. Clark, op. cit., p. 70.

6. National Petroleum Council, *Guide to National Petroleum Council Report on US Energy Outlook* (Washington, DC: 1973), p. 5.

7. Jerome Weingart, 'Surviving the Energy Crunch,' *Environmental Quality*, January 1973, pp. 29–33, 67.

8. William Ophuls, *Ecology and the Politics of Scarcity* (San Francisco: Freeman, 1977), p. 87.

9. Statistics from Frank H. Oram, Associate Director, World Population Society, Washington, DC, August 1978.

10. 'Our Population Predicament: A New Look,' from Population Reference Bureau Inc., vol. 34, no. 5, December 1979.

11. Robert L. Heilbroner, 'Boom and Crash,' *The New Yorker*, August 28, 1978, p. 70.

12. William Ophuls, 'The Scarcity Society,' *Skeptic*, July–August 1974, pp. 50–51.

13. Lee Schipper, 'Energy: Global Prospects 1985–2000,' *Bulletin of the Atomic Scientists*, March 1978, p. 58.

14. Ibid.

15. Hobart Rowen, 'Oil Supply Adequate, Possibly to 1990s, Trilateral Commission Study Concludes,' *Washington Post*, June 14, 1978, p. D9.

16. Emile Benoit, 'The Coming Age of Shortages, Part I,' *Bulletin of the Atomic Scientists*, March 1978, p. 9.

17. Lester R. Brown, *The Twenty-ninth Day* (New York: Norton, 1978), pp. 99–100.

18. Benoit, op. cit., p. 9.

19. Steven Rattner, 'Synthetic-Fuel Plans Are Raising Environment and Cost Questions,' *New York Times*, July 6, 1979, p. 1.

20. Julian McCaull, 'Wringing Out the West,' *Environment* 16 (1974): 10.

21. Peter J. Schuyten, 'The Synthetic Solution: The Rub Is in the Cost,' *New York Times*, July 15, 1979, sec. 3, p. 1.

22. Edward Cowan, 'Synthetic Fuel Costs Called High,' *New York Times*, July 11, 1979, p. D7.

23. John M. Berry, 'Goal May Prove Elusive, Cost Too High,' *Washington Post*, July 29, 1979, p. F1.

24. National Academy of Sciences, *Energy and Climate* (Washington, DC: US Government Printing Office, 1977).

25. Worldwatch Institute, 'The Global Environment and Basic Human Needs,' report to the US Council on Environmental Quality, 1978, p. 36.

26. Joanne Omang, 'Synthetic Fuels Danger to Climate, Scientists Say,' *Washington Post*, July 11, 1979, p. A3.

27. Rattner, op. cit.

28. Worldwatch Institute, op. cit., p. 33.

29. David Dickson, 'Nuclear Power Uncconomic Says Congressional Committee,' *Nature*, May 11, 1978, p. 91.

30. Helen Caldicott, *Nuclear Madness* (Brookline, Mass.: Autumn Press, 1978), p. 43.

31. Ibid., p. 51.

32. Ibid., p. 23.

33. Office of Technology Assessment, *The Effects of Nuclear War* (Washington, DC: US Government Printing Office, 1979).

34. Luther J. Carter, 'Radioactive Wastes: Some Urgent Unfinished Business,' *Science*, February 18, 1977, p. 61.

35. Caldicott, op. cit., pp. 56–63.

36. 'Surveying the Radioactive Waste Dilemma: An Overview,' Critical Mass Energy Project, Washington, DC, August 1978, p. 5.

37. W. Jackson Davis, *The Seventh Year: Industrial Civilization in Transition* (New York: Norton, 1979), p. 65.

38. Wilson Clark, *Energy for Survival* (New York: Doubleday/Anchor Books, 1975), p. 320.

39. Malcolm W. Browne, 'Fusion Power: Is There Still an

Eldorado for Energy?' *New York Times*, April 15, 1979, p. E9.

40. Ibid.

41. Amory Lovins, 'A Light on the Soft Energy Path,' in *Sun: A Handbook for the Solar Decade*, ed. Stephen Lyons (San Francisco: Friends of the Earth, 1978).

42. Sam Love, 'The New Look of the Future,' *The Futurist* 11 (April 1977): 80.

43. Richard England and Barry Bluestone, 'Ecology and Social Conflict,' in *Toward a Steady State Economy*, ed. Herman E. Daly (San Francisco: Freeman, 1973), p. 196.

44. Preston Cloud, 'Mineral Resources in Fact and Fancy,' in *Toward a Steady State Economy*, ed. Herman Daly (San Francisco: Freeman, 1973).

45. Ibid.

46. *US Long Term Economic Growth Prospects: Entering a New Era*, Studies for the Joint Economic Committee of the US Congress (Washington, DC: US Government Printing Office, January 25, 1978).

47. Edward Goldsmith, 'Settlements and Social Stability,' in *Alternatives to Growth I*, ed. Dennis Meadows (Cambridge, Mass.: Ballinger, 1977), p. 331.

48. Lester R. Brown, 'Resource Trends and Population Policy: A Time for Reassessment,' Worldwatch Papers, no. 29, May 1979.

49. Benoit, op. cit.

50. Ophuls, *Ecology and the Politics of Scarcity*, op. cit., p. 87.

51. S. L. Blum, 'Tapping Resources in Municipal Solid Waste,' *Science* 191 (1976): 669–75.

52. Dennis Hayes, *Rays of Hope* (New York: Norton, 1977), p. 139.

PART FIVE
Entropy and the Industrial Age

1. Sylvia Porter, 'Acid of Inflation Erodes Nation's Moral Values,' *Washington Post*, September 11, 1978, p. B6.

2. Barry Commoner, *The Poverty of Power* (New York: Bantam Books, 1977), p. 200.

3. Ibid., p. 200.

4. Ibid., p. 201.

5. Ibid.

6. Ibid., pp. 208–9.

7. Ibid., p. 207.

8. Leslie Ellen Nulty, *Understanding the New Inflation: The Importance of Basic Necessities* (Washington, DC: Exploratory Project for Economic Alternatives, 1977), pp. D19–20.

9. Ibid., pp. D26–30.

10. Ibid., p. D13.

11. *Environmental Quality*, Ninth Annual Report of the Council on Environmental Quality (Washington, DC: US Government Printing Office, December 1978), p. 270.

12. Bureau of Labor Statistics, November 1978.

13. Herman Daly, *Steady State Economics* (San Francisco, CA: W. H. Freeman Co., 1977).

14. Herman Daly, 'The Economic Thought of Frederick Soddy,' in *History of Political Economy*, vol. 12, no. 4, (Durham, NC: Duke University Press, 1980), pp. 469–88.

15. Ibid., p. 475.

16. Ibid., p. 475.

17. Dennis Hayes, *Rays of Hope* (New York: Norton, 1977), p. 91.

18. Ibid., p. 101.

19. Orville Schell, 'Inside the Food Technology Bazaar,' *Mother Jones*, February–March 1979.

20. Hayes, op. cit., p. 148.

21. Jackson W. Davis, *The Seventh Year* (New York: Norton, 1979), p. 126.

22. Barbara Ward, *The Home of Man* (New York: Norton, 1976), p. 49.

23. G. Tyler Miller, Jr., *Energetics, Kinetics and Life: An Ecological Approach* (Belmont, Calif.: Wadsworth, 1971), p. 306.

24. *The Works of Jonathan Swift*, ed. Walter Scott (Edinburgh, 1814), 12:176.
25. Wilson Clark, *Energy for Survival* (Garden City, NY: Doubleday/Anchor Books, 1975), p. 169.
26. Ibid., p. 170.
27. Fred Warshofsky, *Doomsday: The Science of Catastrophe* (New York: Pocket Books, 1979), p. 223.
28. *Environmental Quality*, op. cit., p. 270.
29. Peter Farb, *Humankind* (Boston: Houghton, Mifflin, 1978), pp. 181–82.
30. Clark, op. cit., p. 179.
31. Fred Harris, *The New Populism* (New York: Saturday Review Press, 1973), p. 85.
32. *Food: Green Grow the Profits*, ABC television.
33. Ibid.
34. Clark, op. cit., p. 171.
35. Miller, op. cit., p. 300.
36. *Environmental Quality*, op. cit., p. 276.
37. Clark, op. cit., p. 172.
38. *Environmental Quality*, op. cit., p. 276.
39. Ibid., p. 278.
40. Ibid., p. 277.
41. Clark, op. cit., p. 173.
42. *Environmental Quality*, op. cit., p. 274.
43. Lester R. Brown, *The Twenty-ninth Day* (New York: Norton, 1976), p. 49.
44. Clark, op. cit., p. 174.
45. Worldwatch Institute, *The Global Environment and Basic Human Needs*, A Report to the Council on Environmental Quality (Washington, DC, 1978).
46. *Facts and Trends*, 12th ed. (Washington, DC: Transportation Association of America, 1976), p. 3.
47. Emma Rothschild, *Paradise Lost* (New York: Random House, 1973), p. 18.
48. George W. Brown, *The Freeway Failure*, Proceedings of the Third National Conference on the Transportation Crisis, Washington, DC, June 10, 1972, p. 4.
49. Clark, op. cit., p. 162.
50. Ibid., pp. 157–61.

51. Ibid., p. 160.
52. Ibid.
53. Commoner, op. cit., p. 165.
54. A. Q. Mowbray, *Road to Ruin* (Philadelphia: Lippincott, 1969), p. 15.
55. K. R. Schneider, *Autokind v. Mankind* (New York: Schocken, 1972), p. 123.
56. *Facts and Trends*, op. cit., p. 3.
57. Statement of Robert M. Kennan, Jr., 'Hearings on Future Highway Needs,' Subcommittee on Transportation of the House Committee on Public Works, 93rd Cong., 1st Sess., p. 480.
58. Mowbray, op. cit., p. 14.
59. Ibid., p. 15.
60. Helen Leavitt, *Superhighway: Superhoax* (New York: Ballantine, 1971), p. 13.
61. Ibid., pp. 13, 258.
62. Ibid., p. 257.
63. Ibid., p. 265; Mowbray, op. cit., p. 29.
64. Leavitt, op. cit., p. 20.
65. Ibid., pp. 220–63.
66. Clark, op. cit., p. 158.
67. Brown, op. cit., p. 2.
68. Mowbray, op. cit., p. 14.
69. Ibid., p. 12.
70. Leavitt, op. cit., p. 6.
71. Mowbray, op. cit., p. 68.
72. Ibid., p. 33.
73. Robert Goodman, *After the Planners* (New York: Simon & Schuster, 1971), p. 79.
74. Council on Environmental Quality, *The Costs of Sprawl* (Washington, DC: 1974).
75. Mowbray, op. cit., p. 71; Clark, op. cit., p. 110.
76. Harold M. Schmeck, Jr., 'Lower Level Lead Exposure Tied to Child Brain Damage,' *New York Times*, March 29, 1979, p. A18.
77. Ibid.
78. 'Drift Away from Big Cities Goes On,' *US News and World Report*, November 15, 1976.

79. Davis, op. cit., p. 229.
80. Murray Bookchin, *The Limits of the City* (New York: Harper & Row, 1974), p. 92.
81. Kirkpatrick Sale, 'The Polis Perplexity: An Inquiry into the Size of Cities,' *Working Papers*, January–February 1978, p. 75.
82. 'Drift Away from Big Cities Goes On,' op. cit.
83. Sale, op. cit., p. 66.
84. Ward, op. cit., p. 4.
85. Ellen M. Bussey, *The Flight from Rural Poverty* (Lexington, Mass.: Heath, 1973), p. 103.
86. Lewis Mumford, 'The Natural History of Urbanization,' in *Man's Role in Changing the Face of the Earth*, ed. William L. Thomas, Jr. (University of Chicago Press, 1956).
87. Bookchin, op. cit., p. 33.
88. A. Wolman, 'The Metabolism of Cities,' *Scientific American* 213 (1965): 178–90.
89. Mary Thornton, 'Food Exports Threatened by Loss of Farmland in US,' *Washington Star*, November 26, 1979, p. 2.
90. Bookchin, op. cit., p. 72.
91. Woldman, op. cit.
92. Clark, op. cit., p. 197.
93. Ibid.
94. Kirby and Prokopovitsch, 'Technological Insurance Against Shortages in Minerals and Metals,' *Science* 191 (1976): 713–19.
95. Nancy Humphrey et al., *The Future of Cleveland's Capital Plant* and *The Future of New York City's Capital Plant* (Washington DC: Urban Institute, 1979).
96. *Inadvertent Climate Modification*, Sponsored by the Royal Swedish Academy of Sciences and the Royal Swedish Academy of Engineering Science (Cambridge, Mass.: MIT Press, 1971), p. 12.
97. Sale, op. cit., p. 67.
98. Ibid., pp. 68–69.
99. Ward, op. cit., p. 254; Sale, op. cit., p. 70.
100. Neil Seldman, 'Who Takes Out the Garbage in DC?'

Building System Design, October–November 1975; Kathy Sylvester, 'Monumental Garbage Problem,' *Washington Star,* August 14, 1979, p. B1.

101. Sale, op. cit.
102. Leopold Kohr, *Breakdown of Nations* (New York: Rinehart, 1957).
103. *Environmental Quality,* op. cit.; 'Drift Away from Big Cities Goes On,' op. cit.
104. Source: National Defense Clearing House, Washington, DC, 1979.
105. Source: Center for Defense Information, Washington, DC, 1979.
106. Seymour Melman, 'Beating Swords into Subways,' *New York Times Magazine,* November 19, 1978.
107. Ibid.
108. John K. Cooley, 'Oil Crunch Worries US Military,' *Christian Science Monitor,* May 17, 1979, p. 17.
109. Ibid.
110. Tristram Coffin, 'Conversion, the Answer to Inflation and Recession,' *Washington Spectator,* February 1, 1979.
111. Marion Anderson, *The Impact of Military Spending on the Machinists Union,* Washington, DC, International Association of Machinists, January 1979.
112. Ibid.
113. Coffin, op. cit.
114. Source: *The Bulletin of Atomic Scientists,* available from SANE, Washington, DC, 1979.
115. Source: Center for Defense Information, Washington, DC, 1979.
116. Ibid.
117. Coffin, op. cit.
118. John Harper, 'MX Missile Raises Specter of Massive Land Grab,' in *Wilderness Report* (Washington, DC., Wilderness Society, January 1979).
119. Letter from Robert F. Bennett to President Jimmy Carter, October 2, 1978.
120. *Congressional Record,* June 27, 1979, pp. S8740–43.
121. Telephone interview with Earl Ravenal, October 1979.

122. *Your Taxes, Your Choices*, pamphlet issued by Coalition for a New Foreign and Military Policy, Washington, DC, 1979.

123. Source: National Defense Clearing House, 1979.

124. Eugene Schwartz, *Overskill: The Decline of Technology in Modern Civilization* (New York: Quadrangle, 1971), p. 216.

125. Hayes, op. cit., p. 209.

126. *What Could Your Tax Dollars Buy?* (Washington DC: SANE, December 1978).

127. Quote from Nicholas Georgescu-Roegen, in hearings before the Joint Economic Committee, 'Long-Term Economic Growth,' November 9, 1976, p. 15.

128. Henry Adams, *The Degradation of the Democratic Dogma* (New York: Macmillan, 1949).

129. Source: US Department of Commerce, Industry and Trade Administration estimate that in 1979 advertising expenditures will reach $47·23 billion (ITA press release 79-1, January 3, 1979).

130. Jerry Mander, *Four Arguments for the Elimination of Television* (New York: Morrow Quill, 1978).

131. Peter Schrag, *Mind Control* (New York: Pantheon, 1978), pp. 33–34.

132. Ibid., p. 33.

133. Ibid., p. 43.

134. 'Is Anyone Out There Listening,' a CBS News Report, Parts I and II, August 22–23, 1978.

135. Ibid.

136. Ibid.

137. Ibid.

138. Ibid.

139. Ibid.

140. Ibid.

141. Thomas McKeown, 'A Historical Appraisal of the Medical Task,' in *Medical History and Medical Care*, ed. Gordon McLachlan and Thomas McKeown (London: Oxford University Press, 1971), p. 36.

142. *United States Chartbook: Health* (Washington DC: HEW, National Center for Health Statistics, 1976–77), pp. 1–5.

143. Bernard Dixon, *Beyond the Magic Bullet* (New York: Harper & Row, 1978), p. 3.

144. *United States Chartbook*, op. cit.

145. Dixon, op. cit., p. 226.

146. Ivan Illich, *Medical Nemesis* (New York: Bantam, 1977), pp. 18–19.

147. Dixon, op. cit., pp. 72–74.

148. Ted Howard and Jeremy Rifkin, *Who Should Play God?* (New York: Delacorte, 1977), p. 199.

149. Milton Silverman and Philip Lee, *Pills, Profits and Politics* (Berkeley: University of California Press, 1974).

150. Ibid.

151. Illich, op. cit., p. 23.

152. House Subcommittee on Oversight and Investigations, 1976, as reported in the *Washington Star*, April 1977.

153. John B. McKinlay and Sonja M. McKinlay, 'The Questionable Contribution of Medical Measures to the Decline of Mortality in the US in the Twentieth Century,' *Milbank Memorial Fund Quarterly: Health and Society*, Summer 1977, p. 425.

154. John B. McKinlay and Sonja M. McKinlay, 'A Refutation of the Thesis That the Health of the Nation Is Improving,' a study report as part of a research project supported by grants from Milbank Memorial Fund (to Boston University) and the Carnegie Fund (to Radcliffe Institute), pp. 412–14.

155. Michael J. Canlon, 'EPA Cites US Environment as a Leading Death Cause,' *Washington Post*, August 27, 1978.

156. Emile Benoit, 'A Dynamic Equilibrium Economy,' *Bulletin of the Atomic Scientists*, February 1976.

157. *East West Journal*, August 1977, p. 13.

158. Ophuls, op. cit., pp. 78–79.

159. Daly, op. cit., pp. 193–94.

160. Douglas M. Costle, Administrator, 'Defense by Disaster: Proving the Value of Environmental Protection,' EPA Speech, Washington, DC, March 29, 1979.

161. Stewart T. Herman, *The Health Costs of Air Pollution* (Fairfax, Va.: American Lung Association, 1977).

PART SIX
Entropy: A New World View

1. George C. Wilson and Jim Hoagland, 'Army Is Drafting Plans for "Quick Strike" Force,' *Washington Post,* June 22, 1979, p. A-2.
2. Lawrence A. Mayer, 'Climbing Back from Negative Growth,' *Fortune,* August 1975.
3. Emile Benoit, 'The Coming Age of Shortages, Part I,' *Bulletin of the Atomic Scientists,* January 1976, p. 14.
4. The Hunger Project, San Francisco, California, 1978.
5. Anil Agarwal, 'New Strategy for World Health,' *New Scientist,* June 22, 1978, p. 821.
6. C. T. Kurien, 'A Just, Participatory and Sustainable Society: A Third World Perspective,' paper presented at the Conference on Faith, Science and the Future, World Council of Churches, Boston, Mass., July 12–24, 1979.
7. Herman Daly, 'The Ecological and Moral Necessity for Limiting Economic Growth,' paper presented at the Conference on Faith, Science and the Future, World Council of Churches, Boston, Mass., July 12–24, 1979.
8. Graham Hovey, 'More Crowding Forecast in World's Poorer Cities,' *New York Times,* August 16, 1979, p. D1.
9. Boyce Rensberger, 'Expert Says Only Hope to Feed World Is with Food Production Unlike That in US,' *New York Times,* December 8, 1976, p. A19.
10. O. A. El Kholy, 'Science, Technology and the Future: An Arab Perspective,' paper presented at the Conference on Faith, Science and the Future, World Council of Churches, Boston, Mass., July 12–24, 1979.
11. Robert L. Heilbroner, *An Inquiry into the Human Prospect* (New York: Norton, 1974), p. 86.
12. Hobart Rowen, 'Miller Urges "Austerity" to Cut Inflation,' *Washington Post,* September 7, 1979, p. 1.
13. For information on various solar techniques, see: Denis Hayes, *Rays of Hope* (New York: Norton, 1977) and Stephen Lyons, ed., *Sun: A Handbook for the Solar Decade* (San Francisco: Friends of the Earth, 1978).

14. Ibid.
15. Source: Resources for the Future, Washington, DC, June 1979.
16. W. Jackson Davis, *The Seventh Year: Industrial Civilization in Transition* (New York: Norton, 1979), p. 69.
17. Wilson Clark, *Energy for Survival* (Garden City, NY: Doubleday/Anchor Books, 1975), p. 115.
18. Tristram Coffin, 'Conversion, the Answer to Inflation and Recession,' *Washington Spectator,* February 1, 1979.
19. E. F. Schumacher, *Good Work* (New York: Harper & Row, 1979), p. 18.
20. Murray Bookchin, 'Technology for Life,' in Lyons, op. cit.
21. William Ophuls, *Ecology and the Politics of Scarcity* (San Francisco: Freeman, 1977).
22. Howard Odum, 'Net Energy from the Sun,' in Lyons, op. cit.
23. Richard Munson, 'Ripping off the Sun,' *The Progressive,* September 1979, p. 13.
24. Hayes, op. cit., p. 20.
25. Donald C. Winston, 'There Goes the Sun,' *Newsweek,* December 3, 1979, p. 35.
26. Nicholas Georgescu-Roegen, 'Technology Assessment: The Case of the Direct Use of Solar Energy,' *Atlantic Economic Journal* 6 (December 1978): 20.
27. Munson, op. cit., p. 12.
28. Ibid., p. 13.
29. Ibid., p. 14.
30. Schumacher, op. cit., p. 123.
31. All quotes in this paragraph from *Less Is More,* ed. Goldian Vanden Broeck (New York: Harper & Row, 1978).
32. Ibid.
33. E. F. Shumacher, *Small Is Beautiful* (New York: Perennial, 1973), p. 54.
34. Hazel Henderson, *Creating Alternative Futures* (Berkeley: Windhover, 1978), p. 394.
35. Mark Satin, *New Age Politics* (New York: Delta Books, 1978), p. 29.

36. Harris poll, *Washington Post*, May 23, 1977.
37. C. W. Hollister, 'Twilight in the West,' in *The Transformation of the Roman World*, ed. L. White, Jr. (Berkeley: University of California Press, 1966), p. 204.
38. Barry Commoner, *The Poverty of Power* (New York: Bantam Books, 1977), p. 163.
39. Sam Love, 'The New Look of the Future,' *The Futurist*, April 1977, p. 78.
40. Schumacher, *Small Is Beautiful*, op. cit., p. 54.
41. Davis, op. cit., pp. 217–18.
42. Nicholas Wade, 'Nicholas Georgescu-Roegen: Entropy the Measure of Economic Man,' *Science*, October 31, 1975.
43. Bertrand Russell, *The Scientific Outlook* (New York: Norton, 1962), p. 85.
44. John Lukacs, *The Passing of the Modern Age* (New York: Harper Torchbooks, 1970), p. 152.
45. Russell, op. cit., pp. 92–93.
46. Ibid.
47. Max Born, *The Restless Universe* (New York: Harper, 1936), p. 277.
48. Will Lepkowski, 'The Social Thermodynamics of Ilya Prigogine,' *Chemical & Engineering News*, April 16, 1979, p. 33.
49. George Vecsey, 'Buddhism in America,' *New York Times Magazine*, June 3, 1979, p. 30.
50. The Gallup Opinion Index, 'Religion in America,' 1977–78.
51. Richard Quebedeaux, *The Young Evangelicals* (New York: Harper & Row, 1974), pp. 127–28.
52. Henlee H. Barnett, *The Church and the Ecological Crisis* (Grand Rapids, Mich.: Erdmans, 1972), p. 69. Also, Francis A. Schaeffer, *Pollution and the Death of Man: The Christian View of Ecology* (Wheaton, Ill.: Tyndale House, 1970), p. 37.
53. Barnett, ibid., pp. 78–79.
54. Schaeffer, op. cit., pp. 91–92.
55. Barnett, op. cit., p. 81.
56. Schaeffer, op. cit., pp. 49–50.

57. Francis A. Schaeffer, *How Should We Then Live?* (Old Tappan, N.J.: Fleming Revell, 1976).
58. Ibid., p. 227.
59. Wendell Berry, *The Unsettling of America* (New York: Avon, 1977), p. 20.
60. 'Entropy Concept in Philosophy,' in *Entropy and Information in Science and Philosophy*, ed. Libor Kubat and Jiri Zeman (New York: Elsevier, 1975), p. 240.

Bibliography

PART ONE
World Views

Adams, Henry. *The Degradation of the Democratic Dogma.* New York: Macmillan, 1919.

Bacon, Francis. *Novum Organum.* Book 1, Aphorisms 2, 71.

Barrett, William. *The Illusion of Technique.* Garden City, NY: Doubleday/Anchor, 1976.

Bluhm, William. *Ideologies and Attitudes.* Englewood Cliffs, NJ: Prentice-Hall, 1974.

Boulding, Kenneth. *The Meaning of the Twentieth Century.* New York: Harper & Row, 1965.

Brinton, Crane. *The Shaping of Modern Thought.* Englewood Cliffs, NJ: Prentice-Hall, 1963.

Bury, J. B. *The Idea of Progress: An Inquiry into its Origin and Growth.* New York: Macmillan, 1932.

Descartes, René. *Discourse on Method.* Part 6.

Ferkiss, Victor. *The Future of Technological Civilization.* New York: Braziller, 1974.

Gardiner, Patrick (ed.). *Theories of History.* New York: Free Press, 1959.

Germino, Dante L. *Modern Western Political Thought.* Chicago: Rand McNally, 1972.

Girvetz, Harry K. *The Evolution of Liberalism.* New York: Collier, 1967.

Himmelfarb, Gertrude. *Darwin and the Darwinian Revolution.* New York: Norton, 1968.

Hofstadter, Richard. *Social Darwinism in American Thought.* Boston: Beacon, 1955.

Holton, Gerald. *Thematic Origins of Scientific Thought: Kepler to Einstein.* Cambridge, Mass.: Harvard University Press, 1973.

Humpshire, Stuart. *The Age of Reason: The 17th Century Philosophers.* New York: New American Library, 1956.

311

Kuhn, Thomas S. *The Structure of Scientific Revolutions*. Chicago: University of Chicago Press, 1970.

Lewis, Elwart (ed.). *Medieval Political Ideas*. New York: Knopf, 1954.

Lindsay, Robert Bruce. *The Role of Science in Civilization*. New York: Harper & Row, 1963.

Lukacs, John. *The Passing of the Modern Age*. New York: Harper, 1970.

Marty, Martin. *A Short History of Christianity*. New York: Collins, World, 1959.

Mumford, Lewis. *Technics and Civilization*. New York: Harcourt, Brace, 1934.

Nisbet, Robert. *Twilight of Authority*. New York: Oxford University Press, 1975.

Pollard, Sidney. *The Idea of Progress*. New York: Basic Books, 1968.

Pynchon, Thomas. 'Entropy.' *Kenyon Review* 22 (Spring 1960).

Randall, John Herman. *The Making of the Modern Mind*. New York: Columbia University Press, 1940.

Roszak, Theodore. *Where the Wasteland Ends*. Garden City, NY: Doubleday/Anchor Books, 1973.

Rousseau, Jean Jacques. *The Social Contract*. Harmondsworth. Middlesex, England: Penguin Books, 1968.

Schrag, Peter. *The End of the American Future*. New York: Simon & Schuster, 1973.

Stent, Gunther S. *The Coming of the Golden Age: A View of the End of Progress*. Garden City, NY: Natural History Press, 1969.

Tawney, R. H. *Religion and the Rise of Capitalism*. New York: Harcourt, Brace, 1926.

Tillinghast, Pardon E. (ed.). *Approaches to History*. Englewood Cliffs, NJ: Prentice-Hall, 1963.

Whitehead, Alfred North. *Science and the Modern World*. New York: Macmillan, 1925.

Angrist, Stanley W., and Loren G. Hepler. 'Demons, Poetry and Life: A Thermodynamic View.' *Texas Quarterly* 10 (September 1967).

Asimov, Isaac. *A Short History of Biology.* Garden City, NY: Natural History Press, 1964.

Asimov, Isaac. 'In the Game of Energy and Thermodynamics You Can't Even Break Even.' *Smithsonian*, August 1970.

Asimov, Isaac. 'What is Entropy?' *Science Digest* 73 (January 1973).

Berry, R. Stephen. 'Recycling, Thermodynamics and Environmental Thrift.' *Bulletin of Atomic Scientists* 28 (May 1972).

Blum, Harold F. *Time's Arrow and Evolution.* Princeton, NJ: Princeton University Press, 1968.

Brillouin, L. 'Life, Thermodynamics and Cybernetics.' *American Scientist* 37 (October 1949).

Capek, Milic. *The Philosophical Impact of Contemporary Physics.* New York: Van Nostrand, 1961.

Daly, Herman E. *Toward a Steady State Economy.* San Francisco: Freeman, 1973.

Davies, Paul. *The Runaway Universe.* New York: Harper & Row, 1978.

Ditta, Mahadev. 'A Hundred Years of Entropy.' *Physics Today* 21 (January 1968).

Dobzhansky, Theodosius. *The Biological Basis of Human Freedom.* New York: Columbia University Press, 1956.

Eddington, Sir Arthur. *The Nature of the Physical World.* Ann Arbor: Ann Arbor Paperbacks, University of Michigan Press, 1958.

Georgescu-Roegen, Nicholas. *Energy and Economic Myths.* Elmsford, NY: Pergamon Press, 1977.

Georgescu-Roegen, Nicholas. *The Entropy Law and the Economic Process.* Cambridge, Mass.: Harvard University Press, 1971.

Georgescu-Roegen, Nicholas. *Analytical Economics: Issues and Problems.* Cambridge, Mass.: Harvard University Press, 1966.

Georgescu-Roegen, Nicholas. 'The Steady State and Ecological Salvation: A Thermodynamic Analysis.' *Bioscience* 27 (April 1977).

Heilbroner, Robert L. *An Inquiry into the Human Prospect.* New York: Norton, 1974.

Hiebert, Erwin N. 'The Uses and Abuses of Thermodynamics in Religion.' *Daedalus* 95 (Fall 1966).

Houston, Jean. 'Prometheus Rebound: An Inquiry into Technological Growth and Psychological Change.' In *Alternatives to Growth I.* Edited by Dennis L. Meadows. Cambridge, Mass.: Ballinger, 1977.

Kelley, Hillary Jay. 'Entropy of Knowledge.' *Philosophy of Science* 36 (June 1969).

Kubat, Libor, and Jiri Zeman (eds.). *Entropy and Information in Science and Philosophy.* New York: Elsevier, 1975.

Lindsay, R. B. 'Entropy Consumption and Values in Physical Science.' *American Scientist* 47 (Autumn–September 1959).

Lindsay, Robert Bruce. *The Role of Science in Civilization.* New York: Harper & Row, 1963.

Lotka, Alfred J. 'Contribution to the Energetics of Evolution.' *Proceedings of the National Academy of Sciences* 8 (1922).

Lotka, Alfred J. 'The Law of Evolution as a Maximal Principle.' *Human Biology: A Record of Research* 17 (September 1945).

Mayr, Ernst. 'Evolution.' *Scientific American*, September 1978.

Miller, G. Tyler, Jr. *Energetics, Kinetics and Life: An Ecological Approach.* Belmont, Calif.: Wadsworth, 1971.

Morowitz, Harold J. *Energy Flow in Biology.* New York: Academic Press, 1968.

O'Manique, John. *Energy in Evolution.* New York: Humanities Press, 1969.

Parsegian, V. L. 'Biological Trends within Cosmic Processes.' *Zygon* 8 (September–December 1973).

Polgar, Steven. 'Evolution and the Thermodynamic Imperative.' In *Human Biology*, vol. 33. Detroit, Mich.: Wayne State University Press, 1961.

Porter, George. 'The Laws of Disorder.' *Chemistry.* Reprint 85. Vol. 41 (May–December 1968); vol. 42 (January – February 1969).

Prigogine, Ilya, Gregoire Nicolis, and Agnes Babloyantz. 'Thermodynamics of Evolution.' *Physics Today*, November 1972.

Rodgers, Donald W. 'An Informal History of the First Law of Thermodynamics.' *Chemistry* 49 (December 1976).

Russell, Bertrand. *The Scientific Outlook*. New York: Norton, 1962.

Schrodinger, Erwin. *What Is Life? The Physical Aspect of the Living Cell*. New York: Macmillan, 1947.

Schumacher, E. F. *A Guide for the Perplexed*. New York: Harper & Row, 1977.

Seifert, H. S. 'Can We Decrease Our Entropy.' *American Scientist* 49 (Summer–June 1961).

Simpson, George Gaylord. *The Meaning of Evolution*. New Haven: Yale University Press, 1949.

Stebbing, L. Susan. 'Entropy and Becoming.' *Philosophy and the Physicists*. New York: Dover, 1958.

Wagner, Vern. *The Suspension of Henry Adams: A Study of Manner and Matter*. Detroit, Mich.: Wayne State University Press, 1969.

Warshofsky, Fred. *Doomsday: The Science of Catastrophe*. New York: Pocket Books, 1979.

Watanabe, Satosi. 'The Concept of Time in Modern Physics and Bergson's Pure Duration.' In *Bergson and the Evolution of Physics*. Edited by D. A. Y. Gunter. Knoxville, Tenn.: University of Tennessee Press, 1977.

White, Leslie A. 'Tools, Techniques and Energy.' In *Cultural and Social Anthropology*. Edited by P. Hammond. New York: Macmillan, 1964.

Wiener, Philip P. (ed.). 'Entropy.' In *The Dictionary of the History of Ideas*. Vol. 2. New York: Scribner's, 1973.

PART THREE
Entropy: A New Historical Frame

Adams, Brooks. *The Law of Civilization and Decay*. Freeport, NY: Books for Libraries, 1971.

Adams, Richard Newbold. *Energy and Structure*. Austin, Texas: University of Texas Press, 1975.

Ayres, Eugene. 'The Age of Fossil Fuels.' In *Man's Role in Changing the Face of the Earth*. Edited by William L. Thomas, Jr. Chicago: University of Chicago Press, 1956.

Barrett, William. *The Illusion of Technique*. Garden City, NY: Doubleday/Anchor Books, 1978.

Bates, Marston. 'Process.' In *Man's Role in Changing the Face of the Earth*. Edited by William L. Thomas, Jr. Chicago: University of Chicago Press, 1956.

Bluhm, William T. *Ideologies and Attitudes: Modern Political Culture*. Englewood Cliffs, NJ: Prentice-Hall, 1974.

Boulding, Kenneth. 'The Interplay of Technology and Values: The Emerging Superculture.' In *Values and the Future: The Impact of Technological Change on American Values*. Edited by Kurt Bauer and Nicholas Rescher. New York: Free Press, 1969.

Boulding, Kenneth. *The Meaning of the Twentieth Century*. New York: Harper & Row, 1965.

Childe, V. Gordon. *Man Makes Himself*. New York: New American Library, 1951.

Congress of the United States, Joint Economic Committee. 'Productivity and Technological Change.' In *US Long-Term Economic Growth Prospects: Entering a New Era*. Washington, DC: US Government Printing Office, January 25, 1978.

Daly, Herman E. *Steady-State Economics*. San Francisco: Freeman, 1977.

Darby, H.C. 'The Cleaning of the Woodland.' In *Man's Role in Changing the Face of the Earth*. Edited by William L. Thomas, Jr. Chicago: University of Chicago Press, 1956.

Douglas, Jack D. (ed.). *The Technological Threat*. Englewood Cliffs, NJ: Prentice-Hall, 1971.

Dubos, René. *The Torch of Life*. New York: Simon & Schuster, 1962.

Ellul, Jacques. *The Technological Society*. New York: Random House/Vintage Books, 1964.

Environmental Quality Council. Executive Office of the President. *Environmental Quality, Ninth Annual Report*. Washington, DC: US Government Printing Office, December 1978.

Farb, Peter. *Humankind*. Boston: Houghton Mifflin, 1978.

Ferkiss, Victor. *The Future of Technological Civilization*. New York: Braziller, 1974.

Gall, John. *Systemantics: How Systems Work and Especially How They Fail*. New York: Pocket Books, 1975.

Georgescu-Roegen, Nicholas. *Analytical Economics*. Cambridge, Mass.: Harvard University Press, 1966.

Georgescu-Roegen, Nicholas. *The Entropy Law and the Economic Process*. Cambridge, Mass.: Harvard University Press, 1971.

Giarini, Orio, and Henri Louberge. *The Diminishing Returns of Technology*. Oxford: Pergamon Press, 1978.

Harmon, Willis W. 'The Coming Transformation II.' *The Futurist* 11 (April 1977).

Heilbroner, Robert L. *An Inquiry into the Human Prospect*. New York: Norton, 1974.

Heilbroner, Robert L. 'Do Machines Make History?' In *Technology and Culture*. Chicago: University of Chicago Press, 1967.

Hirsch, Fred. *Social Limits to Growth*. Cambridge, Mass.: Harvard University Press, 1978.

Houston, Jean. 'Prometheus Rebound: An Inquiry into Technological Growth and Psychological Change.' In *Alternatives to Growth I*. Edited by Dennis L. Meadows. Cambridge, Mass.: Ballinger, 1977.

Hoyt, Robert S. *Europe in the Middle Ages*. New York: Harcourt, Brace & World, 1966.

Kariel, Henry S. *Beyond Liberalism, Where Relations Grow*. New York: Harper & Row, 1978.

Kranzberg, Melvin, and Carroll W. Pruell, Jr. *Technology in Western Civilization*. New York: Oxford University Press, 1967.

Lewontin, Richard C. 'Adaption.' *Scientific American*, September 1968.

Lindsay, Robert B. *The Role of Science in Civilization*. New York: Harper & Row, 1963.

Lukacs, John. *The Passing of the Modern Age*. New York: Harper, 1970.

Meadows, Donella. 'Limits to Growth Revisited.' In *Finite*

Resources and the Human Future. Edited by Ian G. Barbour. Minneapolis: Augsburg, 1976.

Mesthene, Emmanual G. *Technological Change*. New York: New American Library, 1970.

Miles, Rufus E., Jr. *Awakening from the American Dream*. New York: Universe, 1976.

Mishan, E. J. *The Economic Growth Debate*. London: George Allen & Unwin, 1977.

Morison, Elting E. *Men, Machines and Modern Times*. Cambridge, Mass.: MIT Press, 1966.

Mumford, Lewis. *Technics and Civilization*. New York: Harcourt, Brace, 1934.

Nichols, Roy F. 'The Dynamic Interpretation of History.' *New England Quarterly*, June 1935.

Oltmans, Willem L. *On Growth*. New York: Capricorn Books, 1974.

Ophuls, William. *Ecology and the Politics of Scarcity*. San Francisco: Freeman, 1977.

Ortega y Gasset, José. *The Revolt of the Masses*. New York: Norton, 1932.

Randall, John, Jr. *The Making of the Modern Mind*. Cambridge, Mass.: Houghton Mifflin, Riverside Press, 1940.

Renshaw, Edward F. 'Productivity.' In *US Economic Growth from 1976 to 1986: Prospects, Problems and Patterns*. Vol. 1. Studies for the Joint Economic Committee. Washington, DC: US Government Printing Office, 1976.

Roszak, Theodore. *Where the Wasteland Ends*. Garden City, NY: Doubleday/Anchor Books, 1972.

Rothman, Harry. *Murderous Providence: A Study of Pollution in Industrial Societies*. New York: Bobbs-Merrill, 1972.

Sauer, Carl O. 'The Agency of Man on the Earth.' In *Man's Role in Changing the Face of the Earth*. Edited by William L. Thomas, Jr. Chicago: University of Chicago Press, 1956.

Schwartz, Eugene S. *Overskill: The Decline of Technology in Modern Civilization*. Chicago: Quadrangle, 1971.

Sears, Paul B. 'The Processes of Environmental Change by Man.' In *Man's Role in Changing the Face of the Earth*. Edited by William L. Thomas, Jr. Chicago: University of Chicago Press, 1956.

Slater, Philip. *The Pursuit of Loneliness: American Culture at the Breaking Point.* Boston: Beacon, 1976.

Spengler, Oswald. *The Decline of the West.* Vol. 1. New York: Knopf, 1926.

Spoehr, Alexander. 'Cultural Differences in the Interpretation of Natural Resources.' In *Man's Role in Changing the Face of the Earth.* Edited by William L. Thomas, Jr. Chicago: University of Chicago Press, 1956.

Stent, Gunther S. *The Coming of the Golden Age: A View of the End of Progress.* Garden City, NY: Natural History Press, 1969.

Teich, Albert H. *Technology and Man's Future.* New York: St Martin's Press, 1972.

Toynbee, Arnold. *A Study of History.* New York: Oxford University Press, 1961.

White, Leslie A. 'Tools, Techniques and Energy.' In *Cultural and Social Anthropology.* Edited by P. Hamond. New York: Macmillan, 1964.

White, Lynn, Jr. 'The Historical Roots of Our Ecological Crisis.' *Science*, March 10, 1967.

Wilkinson, Richard G. *Poverty and Progress.* New York: Praeger, 1973.

Wittfogel, Earl A. 'The Hydraulic Civilizations.' In *Man's Role in Changing the Face of the Earth.* Edited by William L. Thomas, Jr. Chicago: University of Chicago Press, 1956.

PART FOUR
Nonrenewable Energy and the
Approaching Entropy Watershed

Clark, Wilson. *Energy for Survival.* New York: Doubleday/ Anchor Books, 1975.

Commoner, Barry. *The Poverty of Power.* New York: Bantam, 1977.

Davis, W. Jackson. *The Seventh Year: Industrial Civilization in Transition.* New York: Norton, 1979.

Hayes, Dennis. *Rays of Hope.* New York: Norton, 1977.

Lovins, Amory. *Soft Energy Paths.* New York: Ballinger, 1977.

Lyons, Stephen (ed.). *Sun: A Handbook for the Solar Decade.* San Francisco: Friends of the Earth, 1978.

Meadows, Donella, Dennis Meadows, et al. *The Limits to Growth.* New York: Universe Books, 1972.

Nash, Hugh (ed.). *Progress As If Survival Mattered.* San Francisco: Friends of the Earth, 1978.

Odum, Howard T., and Elisabeth C. Odum. *Energy Basis for Man and Nature.* New York: McGraw-Hill, 1976.

Ophuls, William. *Ecology and the Politics of Scarcity.* San Francisco: Freeman, 1977.

PART FIVE
Entropy and the Industrial Age

Adams, Richard Newbold. *Energy and Structure.* Austin, Texas: University of Texas Press, 1975.

Baker, Thelma S. (ed.). *The Urbanization of Man.* Berkeley, Calif.: McCutchan, 1972.

Barber, William J. *A History of Economic Thought.* New York: Penguin Books, 1967.

Barbour, Ian G. (ed.). *Finite Resources and the Human Future.* Minneapolis: Augsburg, 1976.

Benoit, Emile. 'The Coming Age of Shortages, Part I.' *Bulletin of the Atomic Scientists,* January 1976.

Berry, R. Stephen. 'Recycling, Thermodynamics and Environmental Thrift.' *Bulletin of the Atomic Scientists,* May 1972.

Berry, Wendell. *The Unsettling of America: Culture and Agriculture.* New York: Avon, 1977.

Bookchin, Murray. *The Limits of the City.* New York: Harper & Row, 1974.

Boulding, Kenneth. *Beyond Economics.* Ann Arbor, Mich.: Ann Arbor Paperbacks, 1960.

Broun, Hon. George E., Jr. 'Transportation and the Energy Crisis.' *Congressional Record,* July 30, 1973.

Brown, Lester R. *The Twenty-ninth Day.* New York: Norton, 1978.

Caldwell, Malcolm. *The Wealth of Some Nations.* 2nd edition. London: Zed Press Ltd, 1977.

Clark, Wilson. *Energy for Survival*. Garden City, NY: Double-day/Anchor Books, 1975.

Commoner, Barry. *The Poverty of Power*. New York: Bantam Books, 1976.

Daly, Herman E. 'On Economics as a Life Science.' *Journal of Political Economy* 76 (May–June 1968).

Daly, Herman E. *Steady State Economics*. San Francisco: Freeman, 1977.

Davis, W. Jackson. *The Seventh Year: Industrial Civilization in Transition*. New York: Norton, 1979.

Deane, Phyllis. *The Evolution of Economic Ideas*. London: Cambridge University Press, 1978.

Dixon, Bernard. *Beyond the Magic Bullet*. New York: Harper & Row, 1978.

Dubos, René. 'Promises and Hazards of Man's Adaptability.' *Environmental Quality in a Growing Economy*. Essays from the 6th RFF Forum. Baltimore: Johns Hopkins Press, 1966.

Eckholm, Erik. 'Disappearing Species: The Social Challenge.' *Worldwatch Paper*, no. 22 (June 1978).

Environmental Quality Council. Executive Office of the President. *Environmental Quality, Ninth Annual Report*. Washington, DC: US Government Printing Office, December 1978.

Farb, Peter. *Humankind*. Boston: Houghton Mifflin, 1978.

Georgescu-Roegen, Nicholas. *Analytical Economics*. Cambridge, Mass.: Harvard University Press, 1966.

Georgescu-Roegen, Nicholas. *Energy and Economic Myths.'* Elmsford, NY: Pergamon Press, 1977.

Georgescu-Roegen, Nicholas. 'Energy and Economic Myths.' *Southern Economic Journal* 41 (January 1975).

Georgescu-Roegen, Nicholas. *The Entropy Law and the Economic Process.'* Cambridge, Mass.: Harvard University Press, 1971.

Georgescu-Roegen, Nicholas. 'Inequality, Limits and Growth from a Bioeconomic Viewpoint.' *Review of Social Economy* 35 (December 1977).

Georgescu-Roegen, Nicholas. 'The Steady State and Ecological Salvation: A Thermodynamic Analysis.' *Bioscience* 27 (April 1977).

Georgescu-Roegen, Nicholas. 'Technology Assessment: The Case of the Direct Use of Solar Energy.' *Atlantic Economic Journal* 6 (December 1978).

Goodman, Robert. *After the Planners.* New York: Simon & Schuster, 1971.

Harrington, Michael. *The Twilight of Capitalism.* New York: Simon & Schuster, 1976.

Heilbroner, Robert L. *The Economic Transformation of America.* New York: Harcourt Brace Jovanovich, 1977.

Heilbroner, Robert L. *The Worldly Philosophers.* New York: Simon & Schuster, 1953.

Henderson, Hazel. *Creating Alternative Futures: The End of Economics.* New York: Berkeley-Windhover, 1978.

Hoffer, Abram, and Morton Walker. *Ortho-Molecular Nutrition.* New Canaan, Conn.: Keats, 1978.

Illich, Ivan. *Medical Nemesis.* New York: Bantam, 1977.

Ise, John. 'The Theory of Value as Applied to Natural Resources.' *American Economic Review* (St Albans, Vermont) 15 (June 1925).

Jacobs, Jane. *The Death and Life of Great American Cities.* New York: Random House/Vintage Books, 1961.

Jegen, Mary Evelyn, and Bruno Manno (eds.). *The Earth Is the Lord's.* New York: Paulist Press, 1978.

Leavitt, Helen. *Super Highway: Superhoax.* New York: Ballantine Books, 1971.

Lekachman, Robert. *A History of Economic Ideas.* New York: McGraw-Hill, 1959.

McKinlay, John B., and Sonja M. McKinlay. 'The Questionable Contribution of Medical Measures to the Decline of Mortality in the US in the Twentieth Century.' *Milbank Memorial Fund Quarterly: Health and Society,* Summer 1977.

McWhinney, Ian R. 'Medical Knowledge and the Rise of Technology.' *Journal of Medicine and Philosophy,* vol. 3, no. 4 (1978).

Meadows, Donella. 'The World Food Problem: Growth Models and Nongrowth Solutions.' In *Alternatives to Growth I.* Edited by Dennis Meadows. Cambridge, Mass.: Ballinger, 1977.

Mesthene, Emmanuel G. *Technological Change*. New York: New American Library, 1970.

Miles, Rufus. *Awakening from the American Dream*. New York: Universe, 1976.

Mishan, Ezra J. *The Costs of Economic Growth*. New York: Praeger, 1969.

Mowbray, A. Q. *Road to Ruin*. Philadelphia: Lippincott, 1969.

National Wildlife Federation. *The End of the Road*. Washington, DC, 1977.

Office of Technology Assessment, US Congress. *The Effects of Nuclear War*. Washington, DC: US Government Printing Office, 1979.

Oltmans, Willem L. *On Growth*. New York: Capricorn, 1974.

Ophuls, William. *Ecology and the Politics of Scarcity*. San Francisco: Freeman, 1977.

Perelman, Michael. 'Energy, Entropy and Economic Value.' *Australian Economic Papers*, June 1976.

Powles, John. 'The Effect of Health Services on Adult Male Mortality.' In *Ethics in Science and Medicine*. Vol. 5. Elmsford, NY: Pergamon Press, 1978.

Powles, John. 'On the Limitations of Modern Medicine.' In *Science, Medicine, and Man*. Vol. 1. Elmsford, NY: Pergamon Press, 1973.

Rothschild, Emma. *Paradise Lost: The Decline of the Auto Industrial Age*. New York: Random House, 1973.

Schwartz, Eugene. *Overskill: The Decline of Technology in Modern Civilization*. Chicago: Quadrangle, 1971.

Turvey, Ralph. 'Side Effects of Resource Use.' *Environmental Quality in a Growing Economy*. Essays from the 6th RFF Forum, Baltimore: Johns Hopkins Press, 1966.

Ullman, Edward L. 'The Role of Transportation and the Basis for Interaction.' In *Man's Role in Changing the Face of the Earth*. Edited by William L. Thomas, Jr. Chicago: University of Chicago Press, 1956.

Wade, Nicholas. 'Entropy, the Measure of Economic Man.' *Science* 190 (October 31, 1975).

Ward, Barbara. *The Home of Man*. New York: Norton, 1976.

Wilkinson, Richard. *Poverty of Power*. New York: Praeger, 1973.

Wogaman, J. Philip. *The Great Economic Debate.* Philadelphia: Westminster Press, 1977.

PART SIX
Entropy: A New World View

Adams, Henry. 'A Letter to American History Teachers.' In *Degradation of the Democratic Dogma.* New York: Peter Smith, 1949.

Adams, Richard N. *Energy and Structure.* Austin: University of Texas Press, 1975.

Ahlstrom, Sydney E. *A Religious History of the American People.* New Haven: Yale University Press, 1972.

Barnett, Henlee H. *The Church and the Ecological Crisis.* Grand Rapids, Mich.: Erdmans, 1972.

Bell, Daniel. *The Coming of Post Industrial Society.* New York: Basic Books, 1973.

Bellah, Robert N. *The Broken Covenant.* New York: Seabury Press, 1975.

Bellak, Leopold. *Overload: The New Human Condition.* New York: Human Sciences Press, 1975.

Berry, Wendell. *The Unsettling of America: Culture and Agriculture.* New York: Avon Books, 1977.

Bloesch, Donald G. *The Evangelical Renaissance.* Grand Rapids, Mich.: Erdmans, 1973.

Blum, Harold F. 'Order, Negentropy and Evolution.' In *Time's Arrow and Evolution.* Princeton, NJ: Princeton University Press, 1968.

Boulding, Kenneth. 'The Entropy Trap.' In *The Meaning of the Twentieth Century.* New York: Harper & Row, 1965.

Brillouin, L. 'Life, Thermodynamics and Cybernetics.' *American Scientists,* October 1949.

Brush, Stephen G. 'Thermodynamics and History.' *The Graduate Journal,* 1967.

Butler, Samuel. *Erewhon.* New York: Penguin Books, 1977.

Callenbach, Ernest. *Ecotopia.* New York: Bantam, 1975.

Capek, Milic. *The Philosophical Impact of Contemporary Physics.* New York: Van Nostrand, 1961.

CBS Reports. 'Is Anyone Out There Learning.' Parts I, II, III. CBS Television, New York, August 22, 23, 24, 1978.

Davis, W. Jackson. *The Seventh Year: Industrial Civilization in Transition*. New York: Norton, 1979.

Dayton, Donald. *Discovering an Evangelical Heritage*. New York: Harper & Row, 1976.

Eddington, Sir Arthur. *The Nature of the Physical World*. Ann Arbor, Mich.: Ann Arbor Paperbacks, University of Michigan Press, 1958.

Eisenstadt, S. N. *The Protestant Ethic and Modernization*. New York: Basic Books, 1968.

Ellul, Jacques. *The Technological Society*. New York: Random House/Vintage Books, 1964.

Esser, Aristide H. 'Environment and Mental Health.' *Science, Medicine and Man*. Vol. 1. Elmsford, NY: Pergamon Press, 1973.

Geiser, Robert L. *Behavior Mod and the Managed Society*. Boston: Beacon, 1976.

Henderson, Hazel. *Creating Alternative Futures: The End of Economics*. New York: Berkeley-Windhover, 1978.

Henry, Carl. *Contemporary Evangelical Thought: A Survey*. Grand Rapids, Mich.: Baker Book House, 1968.

Hiebert, Erwin N. 'The Uses and Abuses of Thermodynamics in Religion.' *Daedalus*, vol. 95, no. 4 (1966).

Holton, Gerald. *Thematic Origins of Scientific Thought*. Cambridge, Mass.: Harvard University Press, 1973.

Hudson, Winthrop S. *Religion in America*. New York: Scribner's, 1973.

Huxley, Aldous. *The Perennial Philosophy*. New York: Harper, 1970.

Iyer, Raghavan. *The Moral and Political Thought of Mahatma Gandhi*. New York: Oxford University Press, 1973.

Jaki, Stanley L. *The Relevance of Physics*. Chicago: University of Chicago Press, 1966.

Jegen, Mary, and Bruno Manno. *The Earth Is the Lord's*. New York: Paulist Press, 1978.

Kelley, Hillary Jay. 'Entropy of Knowledge.' *Philosophy of Science* 36 (June 1969).

Kuhn, Thomas S. *The Structure of Scientific Revolutions.* Chicago: University of Chicago Press, 1962.

Lambert, Frank L. 'Ontology of Evil.' *Zygon* 3 (June 1968).

Lepkowski, Mil. 'The Social Thermodynamics of Ilya Prigogine.' *Chemical and Engineering News,* April 16, 1979.

Lindsay, Robert B. 'Entropy, Consumption and Values.' *American Scientist* 47 (Autumn–September 1959).

Lindsay, Robert B. 'Science and Communication.' In *The Role of Science in Civilization.* New York: Harper & Row, 1963.

Lipowski, Z. J. 'Sensory and Information Inputs Overload.' *Comprehensive Psychiatry* 16 (May–June 1975).

Lovins, Amory B. *Soft Energy Paths.* Cambridge, Mass.: Ballinger, 1977.

Lukacs, John. *The Passing of the Modern Age.* New York: Harper, 1970.

Marty, Martin E. *A Short History of Christianity.* New York: Collins World, 1959.

Miller, G. Tyler, Jr. 'Entropy, Poetry and Literature.' In *Energetics, Kinetics and Life.* Belmont, Calif.: Wadsworth, 1971.

Morowitz, Harold. 'Order, Information and Entropy.' In *Energy Flow in Biology.* New York: Academic Press, 1968.

Morris, David, and Karl Hess. *Neighborhood Power.* Boston: Beacon, 1975.

Mumford, Lewis. *Technics and Civilization.* New York: Harcourt, Brace, 1934.

O'Manique, John. *Energy in Evolution.* New York: Humanities Press, 1969.

Ostwald, Wilhelm. 'Monism as the Goal of Civilization.' Edited and published by the International Committee of Monism, Hamburg, Germany (1913).

Parsegian, V. L. 'Biological Trends within Cosmic Processes.' *Zygon* 8 (September–December 1973).

Perrin, Noel. *Giving Up the Gun.* Boston: Godine, 1975.

Prigogine, Ilya. 'Thermodynamics of Evolution.' *Physics Today* 25 (November 1972).

Quebedeaux, Richard. *The Young Evangelicals.* New York: Harper & Row, 1974.

Russell, Bertrand. *The Scientific Outlook.* New York: Norton, 1962.

Schaeffer, Francis A. *How Should We Then Live?* Old Tappan, NJ: Revell, 1976.

Schaeffer, Francis A. *Pollution and the Death of Man: The Christian View of Ecology.* Wheaton, Ill.: Tyndale House, 1970.

Schrag, Peter. *Mind Control.* New York: Pantheon, 1978.

Schumacher, E. F. *A Guide for the Perplexed.* New York: Harper & Row, 1977.

Schumacher, E. F. *Good Work.* New York: Harper & Row, 1979.

Schumacher, E. F. *Small is Beautiful.* New York: Perennial, 1973.

Schwartz, Eugene. *Overskill: The Decline of Technology in Modern Civilization.* Chicago: Quadrangle, 1971.

Simpson, George G. 'The Search for an Ethic.' In *The Meaning of Evolution.* New Haven: Yale University Press, 1949.

Slossen, Edwin E. 'Wilhelm Ostwald.' In *Major Prophets of Today.* Freeport, NY: Books for Libraries Press, 1914.

Spillmann, Betty E. *The Logic of Life.* New York: Pantheon, 1973.

Stebbing, L. Susan. 'Entropy and Becoming.' In *Philosophy and the Physicists.* New York: Dover, 1958.

Stent, Gunther S. *The Coming of the Golden Age: A View of the End of Progress.* Garden City, NY: Natural History Press, 1969.

Strizenec, Michal. 'Information and Mental Processes.' In *Entropy and Information in Science and Philosophy.* Edited by Libor Kubat and Jiri Zeman. New York: Elsevier, 1975.

Szumilewicz, Irena. 'The Entropy Concept in Philosophy.' In *Entropy and Information in Science and Philosophy.* Edited by Libor Kubat and Jiri Zeman. New York: Elsevier, 1975.

Tawney, R. H. *Religion and the Rise of Capitalism.* New York: Harcourt, Brace, 1926.

Teich, Albert H. (ed.). *Technology and Man's Future.* New York: St Martin's Press, 1972.

Tillich, Paul. *A History of Christian Thought.* New York: Simon & Schuster, 1967.

Toynbee, Arnold J. *A Study of History*. New York: Oxford University Press, 1961.

Vanden Broeck, Goldian (ed.). *Less is More*. New York: Harper, 1978.

Wade, Nicholas. 'Science and Its Contours: Must Rationality Be Rational?' *Science*, September 13, 1974.

Wallis, Jim. *Agenda for a Biblical People*. New York: Harper & Row, 1976.

Weber, Max. *The Protestant Ethic and the Spirit of Capitalism*. New York: Scribner's, 1958.

White, Lynn, Jr. 'The Historical Roots of Our Ecological Crisis.' *Science*, March 10, 1967.

Zeman, Jiri. 'Information, Knowledge and Time.' In *Entropy and Information in Science and Philosophy*. Edited by Libor Kubat and Jiri Zeman. New York: Elsevier, 1975.

Index

333